I0488968

Burris Numerical System - Expressing numbers as a function of space and time.

Written by Lloyd Dudley Burris

http://time-travel.institute

Table of Contents

List of Figures

List of Abbreviations and Symbols

VCNS	Vector Coordinate Numerical System.
BNS	Burris Numerical System.
Age	How long it takes to code or decode a number.
File size	Size of a file in bytes.
Formatted	An equation in proper professional Mathematical format.
Unformatted	An equation in layman format. (So less educated folks can understand.)
ABS	Absolute Value
Int	Integer

Preface

This book a cross between a acadmic paper and a book. It is about a real numercial system that slowley came into being as I worked on developing a way to store larger numbers as smaller numbers. I started on this actually before the internet came out and before computers became common place in the home. At the time I had never heard of terms like data compression. I barely new anything about encryption and I did not even use encryption. There was no such thing as the hard drive or thumb drive. And, I barely knew who Einstien was let along his work and what he did.

But, through the years and through my journey of developing this numerical system I was to be exposed more and more to computers even becoming very good at using them and I learned more about Einstien. That finally that was enough after so much experimenting to give me the edge to throw me over the top into a world where space, time, and computers all meet and come together.

Into a world where time was a very important element in computer processing that is mainly ignored today. Why is it ignored today. Well because no one knows about it. No one knows how to use time with computer processing the way that I use time with computer processing. Sure there is the clock and network packet timing but that is not what I am talking about.

So, after so much hard work over the course of over 25 years I am finally getting my chance to bring to the world the importance of time in computer processing and how it can be used. This is the first book in a series of books entitled time travel with computers. This first book covers storing and retrievieng information mathematically in time.

I would like to give thanks to my wife Sheila Burris and to my kids Meredith Burris and Amy Burris for their love and support which helped me put my work into writing for the general public.

Abstract

This paper is about a numerical system that was developed from Albert Einstiens work on general relativity. The objective of this numerical system is to express numbers that never get any bigger or smaller than a preset size but the length of the entire number being expressed is a function of time rather than a function of the space needed to represent it.

How that came to be this system uses two numbers one is called the reference point and the other is called the event. As data (any number in any base) is coded into this system the two numbers get bigger. When the numbers store to go past their preset size the plus and minus signs are flipped and now the numbers go down in value as data is coded into these numbers. Again when then number gets as small as it can go the plus and minus signs are flipped again. Now when done coding data no matter how much data there was all the user will have is two small numbers at a preset size. Now, where is the data? Mathematically stored in time. To retrieve the data the numbers go into the decoding equations and the data is mathematically retrieved from time. There is no space limit to the amount of data that can be coded. The only limit is time.

The method to acheive this objective was by the inclusion of a point of reference from one place value in the number to the next place value in the number so that each place value in a number is relative to the one before or after it depending on its point of reference. To do this the entire number is expressed as a set of equations

X = V1 in a base Base – V2 in a Base Unformatted

$$X = V1base - V2base$$

N = (V / X) – int(V/X) Unformatted

$$N=(V/X)-\int(V/X)$$

N = N < Base Unformatted

$N=N<Base$ formatted

For Coding:

V = V +- (((absolute value (V - R)) X (base - 1)) + N) (Unformatted)

$$V=V\pm(((|(V-R)|)\times(Base-1))+N)$$

For Decoding:

V = V +- ((((absolute value(V - R) - N) / Base) X (Base - 1)) + N) (Unformatted)

$$V=V\pm((((|(V-R)|-N)/Base)\times(Base-1))+N)$$

For plotting a code

R1 = V1 +- ((ABS(V2 - V1) - N2) / (base - 1)) is the equation for plotting R. (Unformatted)

$$R1=V1\pm((|(V2-V1)|-N2)/(Base-1))$$

For switching charts

V1 = V1 +- (abs(R2 – R1)) unformatted

$V1=V1\pm(|(R2-R1)|)$ formatted

Distance1 coding = D1 = D1 +- ((ABS(V - R) X (Base - 1)) + N) unformatted

$$Distance1\,coding=D1=D1\pm((|(V-R)|\times(Base-1))+N)$$

Distance2 decoding = D2 = D2 +- ((((ABS(V - R) - N) / Base) X (Base - 1)) + N) unformatted

$$Distance2\,decoding=D2=D2\pm((((|(V-R)|-N)/Base)\times(Base-1))+N)$$

The Burris Numerical System or BNS for short also has been called VCNS in a lot of my earlier work so named for the project that started it. VCNS stands for Vector Coordinate Numerical System.

This number is calulated by using the equations to code a regular number in a specific base to a BNS number in such a way that R can be controlled in a predictable way upon decode.

The BNS number itself is observed one place value at a time. Only the last place value of a BNS number is kept upon completion. By keeping only the last place value represented by V and its reference point R and using the equations to decode the previous place values the length of the entire number can be decoded as long as the code was plotted in such a way as to allow R to be known during the decode.

In conclusion with a method to pick V1 and V2 and calulate R and the equations allowing V to stay the same number of digits or controlle the digit length then the length of a number in a numerical base that can be expressed by BNS is now a function of time and space rather than a function of the size of the number itself. Rather than only a function of space.

Here is a basic program to demonstrate this principle. The compiler to run it can be downloaded at. http://time-travel.institute/qb64.zip

Here the program stores instructions to itself as a binary number in BNS for finding R upon decode rather than plotting R. So, when to change R is the information being store within the program. The instructions aid the program in decoding its BNS number on decode.

In normal reality the instructions would of being a binary file that had no limit as to how big it can grow in bytes but now since the file itself is a BNS number the size of the file is only limited to the amount of the time it takes to to decode it which is close to the same amount of time it took to code the information. When decoding is it only necessary to keep the last place value which is represented by V and R. The place value is used to decode the entire number when a decode is readyf. And, that last place value is only a few bytes of infortional rather than a very large file. On the program when it is running hit d to decode. Hit pause on your computer keyboard to pause and look at program output.

```
Rem VCNS - Vector Coordinate Numerical System invented by Lloyd Dudley Burris.

REM COPYRIGHT C 2002 - 2013
REM Lloyd Dudley Burris
```

```
REM Little Rock, Arkansas
print "This program and VCNS is the intellectual property"
print "of Lloyd Dudley Burris Little Rock, Ar"
print "Copyright C 2002  all rights reserved."
print "Hit enter to continue."
input a$
a$ = ""REM ----------------------Initilize R#-------------------------------------
CLS
DIM r#(255)
r#(1) = 1
REM PRINT r#(1)
pt1# = 5
rf# = 100
count = 2
a# = 0
redoa:
DO
a# = a# + pt1#
test = (a# / 2) - INT(a# / 2)
r#(count) = a#
IF (r#(count) / 2) - INT(r#(count) / 2) = 0 THEN r#(count) = r#(count) + 1
REM PRINT r#(count); count
count = count + 1
REM INPUT a$
REM a$ = INKEY$
IF a$ = "s" THEN STOP
IF count &gt;= 254 THEN EXIT DO
LOOP UNTIL a# = rf#
pt1# = pt1# * 10
rf# = rf# * 10
IF count &gt;= 254 THEN GOTO redob:
IF a# + pt1# &gt; 999999999999999# THEN GOTO redob:
GOTO redoa:
redob:
r#(254) = 950000000000001#
r#(255) = 999999999999999#
REM ----------------------Code P# up.--------------------------------------
range = 237
REM p# = 1000
p# = 100000000000001#
RANDOMIZE TIMER
c = 0
REM file$ = "vcout.txt"
REM OPEN file$ FOR OUTPUT AS #1
repeat:
count = 0
DO
bt = (p# / 2) - INT(p# / 2)
IF bt &gt;= .5 THEN bt = 1
COLOR 2, 0
PRINT , , p#; "+"; bt; r#(range); count
REM WRITE #1, p#, bt, r#(range)
REM a$ = INKEY$
REM INPUT a$
IF a$ = "s" THEN SYSTEM
IF (p# + (p# - r#(range) + 1)) &gt; (r#(range + 1)) THEN c = 1 ELSE c = 0
IF c = 0 THEN p# = p# + (p# - (r#(range)) + 1)
IF c = 1 THEN p# = p# + (p# - (r#(range)) + 2)
IF a$ = "s" THEN STOP
IF p# &gt; (r#(range + 1)) THEN range = range + 1
```

```
count = count + 1
IF range = 254 THEN EXIT DO
LOOP UNTIL p# + (p# - (r#(range)) + 2) &gt;= 999999999999999#
REM CLOSE #1
bt = (p# / 2) - INT(p# / 2)
IF bt &gt;= .5 THEN bt = 1
PRINT , , p#; "+"; bt; r#(range); count
a$ = INKEY$
REM INPUT a$
IF a$ = "s" THEN SYSTEM
IF a$ = "d" THEN GOTO Decodedown:
REM --------------------Code P# down.---------------------------------
range = range + 1
count = 0
DO
bt = (p# / 2) - INT(p# / 2)
IF bt &gt;= .5 THEN bt = 1
COLOR 2, 0
PRINT , , p#; "-"; bt; r#(range); count
REM a$ = INKEY$
REM INPUT a$
REM IF a$ = "s" THEN STOP
REM c = INT(RND * 2) + 0
IF (p# - (r#(range) - p# + 1)) &lt; (r#(range - 1)) THEN c = 1 ELSE c = 0
IF c = 0 THEN p# = p# - ((r#(range)) - p# + 1)
IF c = 1 THEN p# = p# - ((r#(range)) - p# + 2)
IF p# &lt; (r#(range - 1)) THEN range = range - 1
count = count + 1
IF range &lt;= 238 THEN EXIT DO
LOOP UNTIL p# &lt; 150000000000000#
bt = (p# / 2) - INT(p# / 2)
IF bt &gt;= .5 THEN bt = 1
PRINT , , p#; "-"; bt; r#(range); count
a$ = INKEY$
REM INPUT a$
IF a$ = "s" THEN SYSTEM
IF a$ = "d" THEN GOTO decode2:
range = range - 1
GOTO repeat:
REM ---------------------------DECODING AREA------------------------------
REM ---------------------------RESTRICTED AREA:PROGRAM AUTHOR ONLY!!!------
decode1:
REM----------------------------[DECODING + TO -]--------------------------
Decodedown: REM from + to -
DO
bt = (p# / 2) - INT(p# / 2)
IF bt &gt;= .5 THEN bt = 1
IF bt = 1 THEN range = range - 1
COLOR 2, 0
PRINT , , p#; "-"; bt; r#(range)
IF bt = 0 THEN p# = p# - ((p# - r#(range) + 1) / 2)
IF bt = 1 THEN p# = p# - ((p# - r#(range) + 2) / 2)
a$ = INKEY$
REM INPUT a$
IF a$ = "s" THEN STOP
bt = (p# / 2) - INT(p# / 2)
IF bt &gt;= .5 THEN bt = 1
test = (bt = 1) AND range &lt;= 237
IF p# = 100000000000001# THEN GOTO finish:
LOOP UNTIL test = -1
```

```
bt = (p# / 2) - INT(p# / 2)
IF bt &gt;= .5 THEN bt = 1
COLOR 2, 0
PRINT , , p#; "-"; bt; r#(range)
REM ---------------------------[DECODING - TO +]--------------------------
range = range + 1
Decodeup: REM from - to +
decode2:
DO
bt = (p# / 2) - INT(p# / 2)
IF bt &gt;= .5 THEN bt = 1
IF bt = 1 THEN range = range + 1
COLOR 2, 0
PRINT , , p#; "+"; bt; r#(range)
IF bt = 0 THEN p# = p# + ((r#(range) - p# + 1) / 2)
IF bt = 1 THEN p# = p# + ((r#(range) - p# + 2) / 2)
a$ = INKEY$
REM INPUT a$
IF a$ = "s" THEN STOP
bt = (p# / 2) - INT(p# / 2)
IF bt &gt;= .5 THEN bt = 1
test = (bt = 1) AND range &gt;= 255
LOOP UNTIL test = -1
bt = (p# / 2) - INT(p# / 2)
IF bt &gt;= .5 THEN bt = 1
COLOR 2, 0
PRINT , , p#; "+"; bt; r#(range)
range = range - 1
GOTO decode1:
finish:
PRINT , , "DECODE FINISHED!!!!"
print "Time travel forward and backwards complete."
print "Hit Enter to finish."
input a$
SYSTEM
```

Here is a C++ program that demonstrates a BNS coding.

```
#include <time.h>
#include <cstdlib>
#include <iostream>
#include <stdlib.h>
#include <fstream>
#include <sstream>
#include <string>
#include <cmath>
// Note R = R = int((V - R) / base)
// V = V +- (((absolute value (V - R)) X (base - 1)) + N)
// V = V +- ((((absolute value(V - R) - N) / Base) X (Base - 1)) + N)
// declare name space
using namespace std;


// start main
int main (int argc, char *argv[])
{
   int v = 1000;
   int r = 999;
   int dist = 0;
   int count = 0;
   int basenum = 0;
```

```
    int rannum1;

    srand (time(NULL));

    cout << "Please enter the base to use." << "\n";
    cin >> basenum;


// main loop
do
{

    do
    {
// code up
    count++;
    rannum1 = rand() % basenum + 1;
    v = v + (((v - r) * (basenum - 1)) + rannum1);
    cout <<"Number " <<  rannum1 << " V " << v << " R " << r << " V - R " << r - v << "\n";
    dist = v - r;
    if (int(dist/ basenum) - (dist / basenum) == .5)
    {
       dist = dist - 1;
    }
    dist = dist / basenum;
    dist = v - r;
    if (int(dist/ basenum) - (dist / basenum) == .5)
    {
       dist = dist - 1;
    }
    r = r + dist;

// end main loop
    } while(v < 9000);

    cout << " Codingdown" << "\n";
//   system("pause");
    count = 0;
// switch r
    dist = v - r;
    r = v + dist;

    do
    {
// code up
    count++;
    rannum1 = rand() % basenum + 1;
    v = v - (((r - v) * (basenum - 1)) + rannum1);
    cout <<"Number " <<  rannum1 << " V " << v << " R " << r << " V - R " << r - v << "\n";
    dist = r - v;
    if (int(dist/ basenum) - (dist / basenum) == .5)
    {
       dist = dist - 1;
    }
    dist = dist / basenum;
    dist = r - v;
    if (int(dist/ basenum) - (dist / basenum) == .5)
    {
       dist = dist - 1;
    }
    r = r - dist;

// end main loop
    } while(v > 1000);

//   system("pause");
```

```
  dist = r - v;
   r = v - dist;

}while(count < 100000);

}
```

2. INTRODUCTION

2.1 HISTORY OF BNS.

Here is how this all started. Years ago I was sitting in a computer science class at UALR in Little Rock, Arkansas. I think the year is 1991 but don't quote me on that. The instructor was talking about numerical systems. Even at that time I was no stranger to numerical systems. I was first introduced to them at Metro Vo-tech in Little Rock Arkansas in 1986 taking electronics and digital electronics. Anyway to make a long story short at UALR as the instructor was talking I kept asking myself one question? What does it have to take so many digits or place values to store or keep large number? The question why? kept ringing through my head. And, when I worked with scientific notation I again found that I had to lose digits in a large number to express it in a small format. Thus my search for a better way to represent large numbers was started.

So, I set out to invent a new numerical system to store or represent large numbers in a short amount of space without loosing digits in the number. I first started working with extremely large numerical systems using hieroglyphic's representing digits place values in different numerical bases higher than base 10. I used everything from regular numbers, to ASCII code to, cusomized fonts to represent higher number bases. Well problems crept in. While these systems in higher bases were fun to work with they did not meet my objectives and goals in one way or another.

Later I worked with with repeating place values where each new place value could represent itself and all the place values below it then I got the ideal: “Why can I not use math equations and regular numerical systems principles to do the same work that I did with my early work on numerical systems? Thus, I started using numerical systems governed by math equations.

Using math equations to govern numerical systems I discovered it was possible to give a numerical system different mathematical and numerical properties different from other numerical systems. By endowing a numerical system with properties now I could change the way that system represents numbers. It was the perfect angle I was looking for.

The Numerical system example I worked with(cut off the fraction part of the numbers.)

Here is an example of where a place value represents itself and all the place values below it.

Note - This is not the BNS number system this paper is about. This one of my early numerical systems governed by equations.

Figure 1

N = the number being encoded in a numerical base.

t = position at which place value was encoded.

T1 = 1 X N + T2 = 16 * N + T3 = 256 * N = V at T3 Coding Unformatted

$$T1 = 1 \times N \ plus \ T2 = 16 \times N \ plus \ T3 = 256 \times N = V \ at \ t3 \ Coding$$

T3 = int(V/256) Decoding Unformatted

$$t3 = \int (v/256) \ Decoding$$

T2 = V - int(V/256) / 16 Decoding Unformatted

$$t2 = v - \int (v/256)/16 \ Decoding$$

T1 = V - int(V/256) - int(V/16) / 1 Decoding Unformatted

$$t1 = v - \int (v/256) - \int (v/16)/1 \ Decoding$$

We are going to code 357 base 10

V = (1 X 3) + (16 X 5) + (256 X 7) Unformatted

$$V = (1 \times 3) + (16 \times 5) + (256 \times 7)$$

V = 1875 Unformatted

$V=1875$

Decoding V

int(1875 / 256) = 7 Unformatted

$\int(1875/256)=7$

V = 1875 - (256 X 7) Unformatted

$V=1875-(256\times7)$

V = 83 Unformatted

$V=83$

int(83/ 16) = 5 Unformatted

$\int(83/16)=5$

V = 83 - (5 X 16) Unformatted

$V=83-(5\times16)$

V = 3 Unformatted

$V=3$

int(3 / 1) = 3 Unformatted

$\int(3/1)=3$

V = 3 - (1 X 3) Unformatted

$V=3-(1\times3)$

V = 0 Unformatted

$V=0$

Now I was doing the same work with numbers and math equations that I was doing in my earlier numerical systems. Well here too problems crept in. My numbers still got longer and longer. I was not saving any space with my numbers and equations. I found myself at a cross roads. I asked myself is my quest in vane? How can I solve such a problem. I got the ideal of using what I called at the time vectors and coordinates to represent numbers because these vectors and coordinates represented a path forward or backwards to its other place values.

Each number would represent a numerical digit in a specific place value but also contain the vector/coordinate to the number before it or after it that represents a numerical number in a numerical place value. These vector/coordinates would be reusable, and could be large or small. And, would never get any bigger or smaller than a specific numerical range.

Well I finally after much work and aggravation I found a way to put vectors/coordinates into a numerical system by revamping my path principle. I can not mathematically store all possible direct paths between numbers as vectors or coordinates but using Einstein's work I borrowed his reference point after reading about his train thought experiment. Now my numbers paths to their other place values in a base are relative to each other. And it works too for storing computer files because computer files can be thought up as just one long number in base 256 because the bytes are from 0 to 255 in a computer file. It fixed my path problem completly.

I called it VCNS after my orginal project name or Burris Numerical System after myself because I am the inventor.

So with vectors and coordinates as I was using each numerical number in a numerical place value is represented one at a time. Each number only goes back to one vector/coordinate but can go forward to the number of vectors/coordinates equal to the numerical base. Each vector/coordinate is expressed as a place value and a time value representing a numerical number in a numerical base. The time value comes in because that is how much time has to pass retreiving/decoding to this number to get to that specific position in the number.

There are more things I discovered about BNS but more about that later.

2.2 Hypothesis

My hypothese is that information can be mathematically stored and retrieved in time using the basic principles of relativity consisting of a reference point and its event in combination with concepts of using numbers as vector/coordinates. Space-time which is composed of information is relative to the observer. Information that is relative to the observer can be expressed with these BNS equations.

Coding with the BNS equation:

V = V + or - ((absolute value(V - R) X (Base - 1)) + N) Unformatted

$$V = V \pm ((|(V-R)| \times (Base-1)) + N)$$

Decoding with the BNS equations:

V = V + or - ((((absolute value(V - R) - N) / Base) X (Base - 1)) + N) Unformatted

$$V = V \pm ((((|(V-R)| - N)/Base) \times (Base-1)) + N)$$

3. METHODS

3.1Demonstrating BNS with Charts and concepts.

In BNS there are two schools of thoughts. The first school of thought helps to understand the concept it is why I kept the name. The concept is a place value in a numerical base representing a number is also a vector/coordinate pointing to its previous place value in a numerical base representing a number. Now that number too is also a vector/coordinate pointing to its previous place value in a numerical base representing a number. This chain repeats until the entire number in a numerical base is represented.

The second school of thought is this. Now, after the first concept is learned understand it seems impossible to keep or store every vector coordinate so I borrowed Albert Einsteins concept of a point of reference from his train thought experiment. Now every place value in a numerical base representing a number is relative to its previous place value in a numerical base representing a number by way of its reference point.

Now since I have a point of reference I can also measure time as in every place value being t1, t2, t3, and so on. This tells not only how long the number is but also how long it will take to get all the place values out when I want them. The place value with a number in a numerical base can always be a preset size because I am using numbers to represent other numbers hince the concept borrowed from vector/coordinte.

So now each place value in a numerical base representing a number is a preset size because of the vector/coordinate concept and at a specific time because of Albert Einsteins point of reference and we only need the very last place value to keep the entire number.

Now, to make all this truly work out since I am now borrowing Albert Einstens work to make my Vector Coordinate Numerical System work out BNS takes on another meaning. To look at numbers as a object moving through space with its reference point. But the object is a vector/coordinate representing a place value in a numerical base with a number that is relative to its previous vector/coordinate representing a place value in a numerical base with a number. This chain repeats until the entire number is expressed.

What I will be doing here is using numbers and equations to represent a numerical system with a numerical base, a place value, and a number in BNS or Burris Numerical System. In the previous context I have just explained.

I need two things here. A reference point and a starting vector/coordinate representing a place value in a numerical base representing a number.

Any number can be used as a reference point but one will do fine for right now. Now that I have reference point I need a vector/coordinate. 1 will also do fine for that as well.

So in figure 2 the the reference point is 1 and the starting vector/coordinate is 1 in base 10.

Figure 2

1 = reference point

1 = 1, 2, 3, 4, 5, 6, 7, 8, 9, 10

0   0 1 2 3 4 5 6 7 8 9

2 = 11, 12, 13, 14, 15, 16, 17, 18, 19, 20

1   0 1 2 3 4 5 6 7 8 9

3 = 21, 22, 23, 24, 25, 26, 27, 28, 29, 30

2   0 1 2 3 4 5 6 7 8 0

Using a chart to express this each number with the same reference point has only one way back to its previous value but goes to 10 choices as to what number it can be next. There is only one path back but the road forks 10 times. We have 10 choices to take going forward because we are using base 10.

so that fig1

1 = 0 or 7 = 6 or 22 = 1 ect...

This is a hard concept to follow. The 7 in chart 1a actually represents a 6 in base 10. 7 is the vector/coordinate with a numercal value of 6.

1 is a vector/coordinate number with a numerical value of 0.

22 is a vector/coordinate number with a numerical value of 1.

I store each number one by one in this system. When I get done I will only have one number in base 10 to represent 1 digit and place value in base ten. I then take that number and decode it through the chart above to get the rest of the digits out. So that at t5 equals a number but t3 equals a different number when decoding.

The number changes as we decode.

Figure 3 use base two binary

1 = reference point

1 = 2 or 3

0 0 1

2 = 4 or 5

0 0 1

3 = 6 or 7

1 0 1

4 = 8 or 9

0 0 1

5 = 10 or 11

1 0 1

6 = 12 or 13

0 0 1

7 = 14 or 15

1 0 1

8 = 16 or 17

0 0 1

9 = 18 or 19

1 0 1

10 = 20 or 21

0 0 1

11 = 22 or 23

1 0 1

12 = 24 or 25

0 0 1

13 = 26 or 27

1 0 1

ect....

Again you can think of this chart like this.

Figure 4

1 = 0 or 1

0 = 0 or 1

1 = 0 or 1

0 = 0 or 1

Here all even numbers = a binary base two 0 and all odd numbers equal a binary base two 1.

so that 11BNS (from the chart) = 110 in binary base two.

Lets start now from the beginning. We are going to store 1010 in binary base two

1r = 2 or 3 but I want a binary one so I go to 3.

3 = 6 or 7 but I want a binary zero so I got o 6.

6 = 12 or 13 but I want a binary one so I go to 13.

13 = 26 or 27 but we want a binary zero so I go to 26.

so 26BNS base 10 = 1010 in a binary base two numerical system.

To decode this number I do this

26 = a binary 0 so I got my zero

26 goes back to a 13.

13 = a binary 1 so I got my 1

13 goes back to a 6.

6 = a binary 0 so I got my zero.

6 goes back to a 3.

3 = a binary 1 so I got my 1.

3 goes back to 1 my reference point so I am done.

26BNSbase10 = 1010 binary base2

Now to start getting advanced. Notice that these vectors storing a binary numerical system go up in value. Will, how am I going to store numbers to infinity if the numbers keeps getting bigger. I would run out of room doing such a thing. Well take a look below.

Say I stored around 18 binary number 1's and 0's in BNS but now I can not store more than 5 digits. Well, I switch my reference point around and go the other direction.

Figure 5

98735 = reference point

98735 = 98734, 98733

98734 = 98732, 98731

98733 = 98730, 98729

ect...

5 = 4, 3.

When I go as low as I can go I switch my reference point around and start going up in value. And, when the numbers get too large I switch the reference point around and make the vector/coordinate numbers go down in value. Also here I used charts to calculate the vectors. The equation

V = V + or - ((ABS(V - R) X (Base - 1)) + N) is the equation for the charts I am showing. Unformated

$V = V + ¿ - ((|(V - R)| \times (Base - 1)) + N)$ is the equation for the charts I am showing.

A = B or C (remember this for the paragraph below)

Just using the distance between A & B can be used to calculate vectors/coordinates because B & C rise twice as fast as a in base two. If this was base 3 it would be 3 times as fast and so forth for each base.

So I count the distance between a & b or a & c. Multiply that by two (base). Then add that back to c or b to get my new vector/coordinate numbers after choosing b or c to go to for a. Please look at the examples below.

Figure 6

A = B or C

1 = 2 or 3 (2 is the choice I will go to here.)

2 - 1 = 1

1 X 2 = 2

2 + 2 = 4, 2 + 3 = 5 .

so I pick 4 or 5 for my next vector/coordinate for my choice of 2.

2 = 4 or 5

Again:

Figure 7

A = B or C

1 = 2 or 3 (3 is the choice I will go to.)

3 - 1 = 2

2 X 2 = 4

2 + 4 = 6, 3 + 4 = 7

so I pick 6 or 7 for my next vector/coordinate for my choice of 3.

3 = 6 or 7

So, getting aquanted with the BNS equation using charts is why I showed this example because BNS is a combination of two schools of thought. The vector/coordinate principle so that we only need the last place value of a number and the point of reference point princple borrowed from Albert Einstein to make every place value in a number in a numerical base with a value relative to each other.

Well, here I can quite easily represent a 1 trillion digit number with 5 digits plus its reference point with BNS but here t(time) = 1 trillion so it would take one trillion calulations with BNS to encode the one trillion digit number into BNS and one trillion calulations to decode the BNS number back out to its orginal format.

Because of the point of reference and time the last place value of the one trillion number would be kept as a vector/coordinate representing a value. This would be the part of the number that represents space per Albert Einsteins work but the rest of the number would be considered mathematically stored in time as time is measured with points of reference per Albert Einstein's work.

So BNS is a numerical system composed of two parts.

1. Space (the vector/coordinate number you can see.)

2. Time (the vector/coordinates number you can not see.) Until their point of referece is reached.

The combination of one and two as the BNS number codes is like looking at information as a object moving through space with its reference point with the arrow of time going forward.

Likewise:

The combination of one and two as the BNS number decodes is like looking at information as a object moving through space with its reference point with the arrow of time going backward.

Putting it all together. Now, here are vectors/coordinates climbing in value and falling in value but R stayed the same. What if R also climbed in value. Noticed the distance between V and R increased every time I encoded a new value. As this distance increases I would find it very difficult if not impossible to switch my reference point around and also to keep things in order. Also, V increases at the rate of V - R + N with N being my next digit that I go to. I can also increment R as well. The key to pandora's box so to speak with non-linear numerical systems is controlling R. R can climb in value, R can fall in value, but R needs to be as close as possible to V to make my numbers work correctly. The best way to control R is to sink it(bring it back into range of V). When decoding I have to know my previous R. Right now im working on knowing the distance between V1 and R2 before I code so that when decoding I will know the distance between V1 and R1 so that a decode can be repeated. Look at the charts below.

Figure 8

14 v1--------7 r1

-                    -

-                    - (21 - 7) = 14 and (24 - 14) = 10

24v2---------21r2

Notice the distance between 24 and 21 is 3. know (3 X 2) + 1 = then distance between 14 and 7. twenty-one 21 will be

the next R for the next decode and the process will complete and decode if needed. Knowing these distances between

V and R and maybe using a little geometery and graphing seems to be the main key to sinking everything in order for coding and decoding. Tables can computed to plan for R as V codes and decodes.

Figure 9

V = 12501, R = 10001 at T1 V - R = 2500

V = 15002, R = 12501 at T2 V - R = 2501

V = 17505, R = 15003 at T3 V - R = 2502

notice the distance between V and R are still increasing. Understanding the figure 9 just put R and V into a chart at each T to get the values figured out.

Each chart with each R is separate charts here.

Figure 10

10001r = 12500v or 12501v

12501r = 15002v or 15003v

15003r = 17504v or 17505v

So, Incrementing R greater than the distance between (A & B) makes the distance go down.

Incrementing R less than the distance between (A & B) makes the distance between V - R go up.

Note - Later in this paper I will give http links for materials to download to help better understand these concepts.

continuing:

Its also possible to use this to control (V - R).

And of course I can make V go down in value by making the reference point greater than V and code V down in value. I can make V go up in value by making the reference point less than V and code V up in value. When decoding the switches will have to be observed to make this work.

The question still remains though, the distance between V - R is still increasing. Hum, how do I fix that. As said I call that sinking my vectors. Changing the distance between V and R at regular or un-regular intervals so that V and R will stay with in fixed limitations of each other as I code and decode data. Notice in a base 2 BNS system A = B or C and B and C increase at a factor of the (Base X 1) of the numerical system being used.

There is more than one way to control R

How to calulate N for vector/coordinates:

1st way I can find N is by dividing V by the X.

N = V/X Unvormatted

$$N=V/X$$

so X equals the distance between vector/coordinates with the same numerical value in the base being used.

1 = 2 or 3

2 = 4 or 5

the distance between 2 & 4 = 2

the distance between 3 & 5 = 2

This mathmatically calculates X to find N.

So X here would be 2 or multiples of 2.

Figure 11

example using base 3.

1 = 2 3 4

2 = 5 6 7

3 = 8 9 10

2 or 5 or 8 divided by 3 = .666.. equals a base three 0

3 or 6 or 9 divided by 3 = 0 equals a base three 1

4 or 7 or 10 divided by 3 = .3333 equals a base three 2

3. METHODS

3.1 DEMONSTRATING BNS WITH EQUATIONS.

BNS is primarily is used by equations not the charts previously shown. The charts I showed were intended to help understand BNS before I get to the equations. I first started with charts in developing my concepts then I made equations from my charts that is how I arrived at the BNS equation. So here goes.

BNS's equation V = V + or - (((absolute value of (v - r) X (base - 1)) + N) Unformatted

$$((|(V-R)| \times (Base-1)) + N)$$
$$VCNS\,'\,s\,equation\,V = V \pm ¿$$

Here I will code binary

Figure 12

Step - binary number coded

----

1. 0

2. 0

3. 0

4. 1

5. 1

6. 1

7. 0

8. 1

9. 1

10. 0

R = 1

V = 1

X = 2

N = 1 for a binary 0

N = 2 for a binary 1

Going up in value.

Step 1

V = V + ((abs(V - R) X (Base - 1)) + N) unformatted

$$V=V+((|(V-R)|\times(Base-1))+N)$$

1 = 1 + ((abs(1 - 1) X (2 - 1)) + 1) unformaatted

$$1=1+((|(1-1)|\times(2-1))+1)$$

V = 2 unformatted

$$V=2$$

N = (V / X) - int(V / X) unformatted

$$N=(V/X)-\int(V/X)$$

N = (2 / 2) – 1 unformatted

$N = (2/2) - 1$

R = 1, V = 2 , N = 0 Which stands for binary 0

Step 2

V = V + ((abs(V - R) X (Base - 1)) + N) unformatted

$V = V + ((|(V - R)| \times (Base - 1)) + N)$

2 = 2 + ((ABS(2 - 1) X (2 - 1)) + 1) unformatted

$2 = 2 + ((|(2 - 1)| \times (2 - 1)) + 1)$

V = 4 unformatted

$V = 4$

N = (V / X) - int(V / X) unformatted

$N = (V / X) - \int (V / X)$

N = (4 / 2) – 2 unformatted

$N = (4/2) - 2$

N = 0 unformatted

$N = 0$

R = 1, V = 4, N = 0 Which stands for binary 0

Step 3

V = V + ((abs(V - R) X (Base - 1)) + N) unformatted

$V = V + ((|(V - R)| \times (Base - 1)) + N)$

4 = 4 + ((ABS(4 - 1) X (Base - 1)) + 1) unformatted

$4 = 4 + ((|(4 - 1)| \times (Base - 1)) + 1)$

V = 8 unformatted

$V = 8$

N = (V / X) - int(V / X) unformatted

$$N = (V / X) - \int (V / X)$$

N = (8 / 2) - int(4 / 2) unformatted

$$N = (8/2) - \int (4/2)$$

N = 0 unformatted

$$N = 0$$

R = 1, V = 8, N = 0 Which stands for Binary 0.

Breaking here in exampe to increase R

V - R

8 - 1 = 7

V - R is increasing to keep up I will increase R.

R = R + 4

1 = 1 + 4

R = 5

Step 4

V = V + ((ABS(V - R) X (Base - 1)) + N) unformatted

$$V = V + ((|(V - R)| \times (Base - 1)) + N)$$

8 = 8 + ((ABS(8 - 5) X (Base - 1)) + 2) unformatted

$$8 = 8 + ((|(8 - 5)| \times (Base - 1)) + 2)$$

V = 13 unformatted

$$V = 13$$

N = (V / X) - int(V / X) unformatted

$$N = (V / X) - \int (V / X)$$

N = (13 / 2) - int(13 / 2) unformatted

$$N = (13/2) - \int (13/2)$$

N = .5 which means binary 1

R = 5, V = 13 N = .5 which stands for binary 1

Breaking here in example to increase R

V - R

13 - 5 = 8 (V - R is getting larger)

R = R + 4

5 = 5 + 4

R = 9

Step 5

V = V + ((ABS(V - R) X (Base - 1)) + N) unformatted

$$V = V + ((|(V - R)| \times (Base - 1)) + N)$$

13 = 13 + ((ABS(13 - 9) X (Base - 1)) + 2) unformatted

$$13 = 13 + ((|(13 - 9)| \times (Base - 1)) + 2)$$

V = 19 unformatted

$$V = 19$$

N = (V / X) - int(V / X) unformatted

$$N = (V / X) - \int (V / X)$$

N = (19 / 2) – int(19/2) unformatted

$$N = (19/2) - \int (19/2)$$

N = .5 which means binary 1

R = 9, V = 19, N = .5 which stands for binary 1

Breaking here in example to increase R

V - R

19 - 9 = 10 (V - R is getting larger)

R = R + 4

9 = 9 + 4

R = 13

Step 6

V = V + ((ABS(V - R) times (Base - 1)) + N) unformatted

$V = V + ((|(V - R)| \times (Base - 1)) + N)$

19 = 19 + ((ABS(19 - 13) X (Base - 1)) + 2) unformatted

$19 = 19 + ((|(19 - 13)| \times (Base - 1)) + 2)$

V = 27 unformatted

$V = 27$

N = (V / X) - int(V / X) unformatted

$N = (V / X) - \int (V / X)$

N = (27 / 2) - int(27 / 2) unformatted

$N = (27/2) - \int (27/2)$

N = .5 which means binary 1

R = 13, V = 27, N = .5 which stands for binary 1

Breaking here in example to increase R

V - R

27 - 13 = 14 (V - R is getting larger)

R = R + 8

13 = 13 + 8

R = 21

Step 7

V = V + ((ABS(V - R) X (Base - 1)) + N) unformatted

$V = V + ((|(V - R)| \times (Base - 1)) + N)$

27 = 27 + ((ABS(27 - 21) X (Base - 1)) + 1) unformatted

$27 = 27 + ((|(27 - 21)| \times (Base - 1)) + 1)$

V = 34 unformatted

$V = 34$

N = (V / X) - int(V / X) unformatted

$N = (V / X) - \int (V / X)$

N = (34 / 2) - int(34 / 2) unformatted

$N = (34/2) - \int (34/2)$

N = 0 which stands for binary 0

R = 21, V = 34, N = 0 which stands for binary 0

Breaking here in example to increase R

V - R

34 - 21 = 13 (V - R is getting smaller)

R = R + 8

21 = 21 + 8

R = 29

Step 8

V = V + ((ABS(V - R) X (Base - 1)) + N) unformatted

$V = V + ((|(V - R)| \times (Base - 1)) + N)$

34 = 34 + ((ABS(34 - 29) X (Base - 1)) + 2) unformatted

$34 = 34 + ((|(34 - 29)| \times (Base - 1)) + 2)$

V = 41 unformatted

$V = 41$

N = (V / X) - int(V / X) unformatted

$N=(V/X)-\int(V/X)$

N = (41 / 2) - int(41 / 2) unformatted

$N=(41/2)-\int(41/2)$

N = .5 which means binary 1

R = 29, V = 41, N = .5 which means binary 1

Breaking here in example to increase R

V - R

41 - 29 = 12 (Here V - R is getting smaller)

R = R + 8

29 = 29 + 8

R = 37

Step 9

V = V + ((ABS(V - R) X (Base - 1)) + N) unformatted

$V=V+((|(V-R)|\times(Base-1))+N)$

41 = 41 + ((ABS(41 - 37) X (Base - 1)) + 2) unformatted

$41=41+((|(41-37)|\times(Base-1))+2)$

V = 47 unformatted

$V=47$

N = (V / X) - int(V / X) unformatted

$N=(V/X)-\int(V/X)$

N = (47 / 2) - int(47/ 2) unformatted

$N=(47/2)-\int(47/2)$

N = .5 which stands for binary 1

R = 37, V = 47, N = .5 which means binary 1

Breaking here in example to increase R

V - R

47 - 37 = 10 (V - R is smaller so it is catching up with V)

R = R + 8

37 = 37 + 8

R = 45 (R has caught up with V here.)

Step 10

V = V + ((ABS(V - R) X (Base - 1)) + N) unformatted

$$V = V + ((|(V - R)| \times (Base - 1)) + N)$$

V =47 + ((ABS(47 - 45) X (Base - 1)) + 1) unformatted

$$V = 47 + ((|(47 - 45)| \times (Base - 1)) + 1)$$

V = 50 unformatted

$$V = 50$$

N = (V / X) - int(V / X) unformatted

$$N = (V / X) - \int (V / X)$$

N = (50 / 2) - int(50 / 2) unformatted

$$N = (50/2) - \int (50/2)$$

N = 0 which means binary 0

Final Values

V = 50, R = 45, N = 0 which means binary 0

So this is what I keep. This is all I keep. This is the number that equal 10 binary values. This is the space part of the number. The rest of the number is mathematically stored in time by way of Albert Einsteins reference points.

V = 50, R = 45, N = 0

With Albert Einsteins reference points

R = The point of reference

Time = The count in the coding or decoding process.

V = The space part of the number.

N = The data at that point in time.

So, mathematically N is stored in time and is relative to the observer. Space-time is relative to the observer. So at time = 10 with a reference point of 45 at V = 50 (V is the space part of the number). The data I have at this point in space-time is a binary 0.

Ok, so now to decode.

Figure 13

The equation for decoding BNS:

V = V + or - ((((ABS(V - R) - N) / Base) X (Base - 1)) + N) unformatted

$$V=V\pm((((|(V-R)|-N)/Base)\times(Base-1))+N)$$

Step 10

V = 50, R = 45, N = 0 or binary 0

V = V + or - ((((ABS(V - R) - N) / Base) X (Base - 1)) + N) unformatted

$$V=V\pm((((|(V-R)|-N)/Base)\times(Base-1))+N)$$

50 = 50 - ((((ABS(50 - 45) - 1) / 2) X (2 - 1)) + 1) unformatted

$$50=50-((((|(50-45)|-1)/2)\times(2-1))+1)$$

V = 47 R = 45 unformatted

$$V=47R=45$$

N = (V / X) - int(V / X) unformatted

$$N=(V/X)-\int(V/X)$$

N = (47 / 2) - int(47 / 2) unformatted

$$N = (47/2) - \int (47/2)$$

N = .5 which a binary 1

I change my R.

R = 45 - 8

R = 37

Final Values

V = 47, R = 37, N = .5 which means binary 1

Step 9

V = 47, R = 37, N = .5 which means binary 1

V = V + or - ((((ABS(V - R) - N) / Base) X (Base - 1)) + N) unformatted

$$V = V \pm ((((|(V-R)| - N)/Base) \times (Base - 1)) + N)$$

47 = 47 - ((((ABS(47 - 37) - 2) / 2) X (2 - 1)) + 2) unformatted

$$47 = 47 - ((((|(47-37)| - 2)/2) \times (2-1)) + 2)$$

V = 41 R = 37 unformatted

$$V = 41\, R = 37$$

N = (V / X) - int(V / X) unformatted

$$N = (V/X) - \int (V/X)$$

N = (41 / 2) - int(41 / 2) unformatted

$$N = (41/2) - \int (41/2)$$

N = .5 which is a binary 1

I change my R.

R = R - 8

R = 37 - 8

R = 29

Final Values

V = 41, R = 29, N = .5 which means binary 1

Step 8

V = 41, R = 29, N = .5 which means binary 1

V = V + or - ((((ABS(V - R) - N) / Base) X (Base - 1)) + N) unformatted

$$V = V \pm ((((|(V - R)| - N)/Base) \times (Base - 1)) + N)$$

V = 41 - ((((ABS(41 - 29) - 2) / 2) X (2 - 1)) + 2) unformatted

$$V = 41 - ((((|(41 - 29)| - 2)/2) \times (2 - 1)) + 2)$$

V = 34, R = 29 unformatted

$$V = 34, R = 29$$

N = (V / X) - int(V / X) unformatted

$$N = (V/X) - \int (V/X)$$

N = (34 / X) - int(34 / 2) unformatted

$$N = (34/X) - \int (34/2)$$

N = 0 which means binary 0

I change my R.

R = R - 8

R = 29 - 8

R = 21

Final Values

V = 34, R = 21, N = 0 which means binary 0

Step 7

V = 34, R = 21, N = 0 which means binary 0

V = V + or - ((((ABS(V - R) - N) / Base) X (Base - 1)) + N) unformatted

$$V = V \pm ((((|(V-R)|-N)/Base)\times(Base-1))+N)$$

V = 34 - ((((ABS(34 - 21) - 1) / 2) X (2 - 1)) + 1) unformatted

$$V = 34 - ((((|(34-21)|-1)/2)\times(2-1))+1)$$

V = 27, R = 21 unformatted

$$V = 27, R = 21$$

N = (V / X) - int(V / X) unformatted

$$N = (V/X) - \int(V/X)$$

N = (27 / 2) - int(27 / 2) unformatted

$$N = (27/2) - \int(27/2)$$

N = .5 which means binary 1

I change my R.

R = R - 8

R = 21 - 8

R = 13

Final Values

V = 27, R = 13, N = .5 which means binary 1

Step 6

V = 27, R = 13, N = .5 which means binary 1

V = V + or - ((((ABS(V - R) - N) / Base) X (Base - 1)) + N) unformatted

$$V = V \pm ((((|(V-R)|-N)/Base)\times(Base-1))+N)$$

V = 27 - ((((ABS(27 - 13) - 2) / 2) X (2 - 1)) + 2) unformatted

$$V = 27 - ((((|(27-13)|-2)/2)\times(2-1))+2)$$

V = 19, R = 13 unformatted

$V=19, R=13$

N = (V / X) - int(V / X) unformatted

$N=(V/X)-\int(V/X)$

N = (19 / 2) - int(19 / 2) unformatted

$N=(19/2)-\int(19/2)$

N = .5 which means binary 1.

I change my R.

R = R - 8

R = 13 - 8

R = 5

Final Values

V = 13, R = 5, N = .5 which means binary 1

Step 5

V = 13, R = 5, N = .5 which means binary 1

V = V + or - ((((ABS(V - R) - N) / Base) X (Base - 1)) + N) unformatted

$V=V\pm((((|(V-R)|-N)/Base)\times(Base-1))+N)$

V = 13 - ((((ABS(13 - 5) - 2) / 2) X (2 - 1)) + 2) unformatted

$V=13-((((|(13-5)|-2)/2)\times(2-1))+2)$

V = 8, R = 5 unformatted

$V=8, R=5$

N = (V / X) - int(V / X) unformatted

$N=(V/X)-\int(V/X)$

N = (8 / 2) - int(8 / 2) unformatted

$N=(8/2)-\int(8/2)$

N = 0 which means binary 0

I change my R.

R = R - 4

R = 5 - 4

R = 1

Final Values

V = 8, R = 1, N = .5 which means binary 0

Step 4

V = 8, R = 1, N = .5 which means binary 0

V = V + or - ((((ABS(V - R) - N) / Base) X (Base - 1)) + N) unformatted

$$V = V \pm ((((|(V - R)| - N) / Base) \times (Base - 1)) + N)$$

V = 8 - ((((ABS(8 - 1) - 1) / 2) X (2 - 1)) + 1) unformatted

$$V = 8 - ((((|(8 - 1)| - 1) / 2) \times (2 - 1)) + 1)$$

V = 4, R = 1 unformatted

$$V = 4, R = 1$$

N = (V / X) - int(V / X) unformatted

$$N = (V / X) - \int (V / X)$$

N = (4 / 2) - int(4 / 2) unformatted

$$N = (4/2) - \int (4/2)$$

N = 0 which means binary 0

I do not change R any more.

Final Values

V = 4, R = 1, N = 0 which means binary 0

Step 3

V = 4, R = 1, N = 0 which means binary 0

V = V + or - ((((ABS(V - R) - N) / Base) X (Base - 1)) + N) unformatted

$$V=V\pm((((|(V-R)|-N)/Base)\times(Base-1))+N)$$

V = 4 - ((((ABS(4 - 1) - 1) / 2) X (2 - 1)) + 1) unformatted

$$V=4-((((|(4-1)|-1)/2)\times(2-1))+1)$$

V = 2, R = 1 unformatted

$$V=2,R=1$$

N = (V / X) - int(V / X) unformatted

$$N=(V/X)-\int(V/X)$$

N = (2 / 2) - int(2 / 2) unformatted

$$N=(2/2)-\int(2/2)$$

N = 0 which means binary 0

I do not change my R

Final Values

V = 2, R = 1, N = 0 which means binary 0

Step 2

V = 2, R = 1, N = 0 which means binary 0

V = V + or - ((((ABS(V - R) - N) / Base) X (Base - 1)) + N) unformatted

$$V=V\pm((((|(V-R)|-N)/Base)\times(Base-1))+N)$$

V = 2 - ((((ABS(2 - 1) - 1) / 2) X (2 - 1)) + 1) unformatted

$$V=2-((((|(2-1)|-1)/2)\times(2-1))+1)$$

V = 1, R = 1, N = Not computed because N not used here.

Decode completed

3. METHODS

3.3 PLOTING A CODE AND DECODE

With BNS/Burris Numerical System to code large numbers or computer files into BNS/Burris Numerical System both R and V may be planned for in advance and a system put into place so that R is always known not only for coding but also for decoding.

If coding from negative to positive and from positive to negative the plus and minus signs in the equations must be flipped so that V will go up in value and down in value but stay within a fixed range. To do that a range for how large or small V can get must be established. R is established within a specfic range of V.

I will be planning a decode and a code of V at the same time. When planning a code or decode of V it must be established that formost the hurdle to over come is the distance between V and R and how to always know the distance from R1 and R2. Do accomplish these goes then BNS must become a distance problem and handled as such.

I will code and decode the number 0010.

Figure 14

The equations I will be using.

Coding

V = V + or - ((ABS(V - R) X (Base - 1)) + N) unformatted

$$V=V\pm((|(V-R)|\times(Base-1))+N)$$

Decoding

V = V + or - ((((ABS(V - R) - N) / Base) X (Base - 1)) + N) unformatted

$$V=V\pm((((|(V-R)|-N)/Base)\times(Base-1))+N)$$

Now two new equations

Distance1 coding = D1 = D1 +- ((ABS(V - R) X (Base - 1)) + N) unformatted

$Distance1\,coding = D1 = D1 \pm ((|(V - R)| \times (Base - 1)) + N)$

Distance2 decoding = D2 = D2 +- ((((ABS(V - R) - N) / Base) X (Base - 1)) + N) unformatted

$Distance2\,decoding = D2 = D2 \pm ((((|(V - R)| - N)/Base) \times (Base - 1)) + N)$

Distance3 = V1 - V2

Distance4 = R1 - R2

When planning or plotting a code and a decode I plan the distances between V - R as it codes up and codes down. I plan distances between V1 - V2, and so forth. I plan distances between R1 - R2, and so forth.

I define distance as V - R or V1 - V2 or R1 - R2.

Specific R's and V's are not known yet.

I plan when and were I will increase or decrease R.

I plan when I will switch the plus and minus signs around in my equations in doing so I also switch V and R around in my equations so my numbers come out positive.

So, after my plan I have a list of distances for V - R. Now I find my V's and R's to fit those distances in my plan so when I work out my equations and everything I code numbers/data with V and R everything comes out ok for both coding and decoding.

So, to make a long story short I need a code plan and a decode plan based on the distances between the variables in my equations. The code/decode plan becomes a distance problem to solve.

Here I will code 0, 1, 1, 0

Figure 15

V - R = 0

Distance1 coding = ((ABS(V - R) X (Base - 1)) + N) unformatted

$Distance1\,coding = ((|(V - R)| \times (Base - 1)) + N)$

D1 = 0 + (((0) X (1)) + 1) unformatted

$$D1 = 0 + (((0) \times (1)) + 1)$$

D1 = 1 N = 0 binary 0

D2= 1 + (((1) X (1)) + 2) unformatted

$$D2 = 1 + (((1) \times (1)) + 2)$$

D2 = 4 N = .5 binary 1

D3 = 4 + (((4) X (1)) + 2) unformatted

$$D3 = 4 + (((4) \times (1)) + 2)$$

D3 = 10 N = .5 binary 1

D4 = 10 + (((10) X (1)) + 1) unformatted

$$D4 = 10 + (((10) \times (1)) + 1)$$

D4 = 21

To show this works.

Figure 16

V = 1 R = 1 unformatted

$$V = 1 \; R = 1$$

V = V +- ((ABS(V - R) X (Base - 1)) + N) unformatted

$$V = V \pm ((|(V - R)| \times (Base - 1)) + N)$$

V = 1 + (((1 - 1) X (2 - 1)) + 1) unformatted

$$V = 1 + (((1 - 1) \times (2 - 1)) + 1)$$

V = 2 R = 1 N = 0 V - R = 1 D1 = 1 unformatted

$$V = 2 \; R = 1 \; N = 0 \; V - R = 1 \; D1 = 1$$

V = V +- ((ABS(V - R) X (Base - 1)) + N) unformatted

$$V = V \pm ((|(V - R)| \times (Base - 1)) + N)$$

V = 2 + (((2-1) X (2 - 1)) + 2) unformatted

$$V = 2 + (((2 - 1) \times (2 - 1)) + 2)$$

V = 5 R = 1 N = .5 V - R = 4 D2 = 4 unformatted

$$V=5\,R=1\,N=.5\,V-R=4\,D2=4$$

V = V + (((V - R) X (Base - 1)) + N) unformatted

$$V=V+(((V-R)\times(Base-1))+N)$$

V = 5 + (((5 - 1) X (2 - 1)) + 2) unformatted

$$V=5+(((5-1)\times(2-1))+2)$$

V = 11 R = 1 N = .5 V - R = 10 D3 = 10 unformatted

$$V=11\,R=1\,N=.5\,V-R=10\,D3=10$$

V = V + (((V - R) X (Base - 1)) + N) unformatted

$$V=V+(((V-R)\times(Base-1))+N)$$

V = 11 + (((11 - 1) X (2 - 1)) + 1) unformatted

$$V=11+(((11-1)\times(2-1))+1)$$

V = 22 R = 1 N = 0 V - R = 21 D4 = 21 unformatted

$$V=22\,R=1\,N=0\,V-R=21\,D4=21$$

But now since I have my distance plan I can pick a orbitrary V Calulate my R where N = 0 and do a test decode.

My coding up plan would allow V - R to get larger till my range is met then V - R is reset to a smaller value and allow my V - R to get larger again. Knowing my plan when I decode I would allow my V - R to get smaller to my limit is reached then reset my V - R to a larger value and repeat till I get all my numbers/file back out.

Demonstrating a decode plan with distance.

Figure 17

D4 = 21 N = 0 binary 0 unformatted

$$D4=21\,N=0\,binary\,0$$

Distance decoding = D = D +- ((((ABS(V - R) - N) / Base) X (Base - 1)) + N) unformatted

$Distance\,decoding = D = D \pm (((((|(V-R)|-N)/Base)\times(Base-1)) + N)$

D3 = 21 - (((((21) - 1) / 2) X (2 - 1)) + 1) unformatted

$D3 = 21 - (((((21)-1)/2)\times(2-1)) + 1)$

D3 = 10 N = .5 binary 1 unformatted

$D3 = 10\,N = .5\,binary\,1$

D2 = 10 - (((((10) - 2) / 2) X (2 - 1)) + 2) unformatted

$D2 = 10 - (((((10)-2)/2)\times(2-1)) + 2)$

D2 = 4 N = .5 binary 1 unformatted

$D2 = 4\,N = .5\,binary\,1$

D1 = 4 - (((((4) - 2) / 2) X (2 - 1)) + 2) unformatted

$D1 = 4 - (((((4)-2)/2)\times(2-1)) + 2)$

D1 = 1 N = 0 binary 0 imfpr,atted

$D1 = 1\,N = 0\,binary\,0$

To show this works.

Figure 18

V = 22 R = 1 N = 0 V - R = 21 D4 = 21 unformatted

$V = 22\,R = 1\,N = 0\,V - R = 21\,D4 = 21$

V = V + or - (((((ABS(V - R) - N) / Base) X (Base - 1)) + N) unformatted

$((((|(V-R)|-N)/Base)\times(Base-1)) + N)$
$V = V \pm ¿$

V = 22 - (((((V - R) - 1) / 2) X (2 - 1)) + 1) unformatted

$V = 22 - (((((V-R)-1)/2)\times(2-1)) + 1)$

V = 11 R = 1 V - R = 10 D3 = 10 N = .5 binary 1 unformatted

$V = 11\,R = 1\,V - R = 10\,D3 = 10\,N = .5\,binary\,1$

V = V + or - (((((ABS(V - R) - N) / Base) X (Base - 1)) + N) unformatted

$$((((|(V-R)|-N)/Base)\times(Base-1))+N)$$
$$V=V\pm¿$$

V = 11 - (((((11 - 1) - 2) / 2) X (2 - 1)) + 2) unformatted

$$V=11-(((((11-1)-2)/2)\times(2-1))+2)$$

V = 5 R = 1 V - R = 4 D2 = 4 N = .5 binary 1 unformatted

$$V=5\,R=1\,V-R=4\,D2=4\,N=.5\,binary\,1$$

V = V + or - (((((ABS(V - R) - N) / Base) X (Base - 1)) + N) unformatted

$$((((|(V-R)|-N)/Base)\times(Base-1))+N)$$
$$V=V\pm¿$$

V = 5 - (((((5 - 1) - 2) / 2) X (2 - 1)) + 2) unformatted

$$V=5-(((((5-1)-2)/2)\times(2-1))+2)$$

V = 2 R = 1 V - R = 1 N = 0 binary 0 D1 = 1 unformatted

$$V=2\,R=1\,V-R=1\,N=0\,binary\,0\,D1=1$$

V = V + or - (((((absolute value(V - R) - N) / Base) X (Base - 1)) + N) unformatted

$$((((|(V-R)|-N)/Base)\times(Base-1))+N)$$
$$V=V\pm¿$$

V = 2 - (((((2 - 1) - 1) / 2) X (2 - 1)) + 1) unformatted

$$V=2-(((((2-1)-1)/2)\times(2-1))+1)$$

V = 1 R = 1 N = No value unformatted

$$V=1\,R=1\,N=No value$$

So, Taking my data I plan out ahead of time what I will know upon decode to always find my R. V is always known as long as R is correct. So I have to know when to change R and what to change R too. This is done by planning out ahead of time my code or decode with distance between V and R at each step of the coding process and what I will have for R.

Here a code will be plotted going backward. I will code the binary number 0110 using distances between V and R.

When calulating distance the value of R needs to be considered because I will be moving R around here. My R is a odd number. So it has to be moved in multiples of at least 2 or more.

Figure 19

So if V = 202 and R = 199 then V - R = 2. If R suddenly becomes 197 then V - R = 4.

So V = 202, R = 199, V - R = 4

V = V + or - ((((ABS(V - R) - N) / Base) X (Base - 1)) + N) unformatted

$$V=V\pm((((|(V-R)|-N)/Base)\times(Base-1))+N)$$

V = 202 - ((((ABS(202 - 199) - 1) / 2) X (2 - 1)) + 1) unformatted

$$V=202-((((|(202-199)|-1)/2)X(2-1))+1)$$

V = 200. (NO this wont do. (I want a binary 1). unformatted

$V=200$ (No this wont do. (I want a binary 1).

N = (V / X) - int(V / X) unformatted

$$N=(V/X)-\int(V/X)$$

N = (200 / 2) - int(200 / 2) unformatted

$$N=(200/2)-\int(200/2)$$

N = 0 (Wrong answer) Again changing R to 197. unformatted

$N=0(Wrong\ answer)\ Again\ changing\ R¿197.$

Changing R from 199 to 197.

So V = 202, R = 197, V - R = 5

V = V + or - ((((ABS(V - R) - N) / Base) X (Base - 1)) + N) unformatted

$$V=V\pm((((|(V-R)|-N)/Base)\times(Base-1))+N)$$

V = 202 - ((((ABS(202 - 197) - 1) / 2) X (2 - 1)) + 1) unformatted

$V = 202 - ((((|(202 - 197)| - 1)/2) \times (2-1)) + 1)$

V = 199, R = 197, V - R = 2 (Good I went from a binary 0 to a binary 1)

Changing R to 195

V = 199, R = 195, V - R = 4

V = V + or - ((((ABS(V - R) - N) / Base) X (Base - 1)) + N) unformatted

$V = V \pm ((((|(V - R)| - N)/Base) \times (Base - 1)) + N)$

V = 199 - ((((ABS(199 - 195) - 2) / 2) X (2 - 1)) + 2) unformatted

$V = 199 - ((((|(199 - 195)| - 2)/2) \times (2-1)) + 2)$

V = 196

N = (V / X) - int(V / X) unformatted

$N = (V/X) - \int (V/X)$

N = (196 / 2) - int(196 / 2) unformatted

$N = (196/2) - \int (196/2)$

N = 0 No this will not do.

Changing R to 193.

V = 199, R = 193, V - R = 6

V = 199 - ((((ABS(199 - 193) - 2) / 2) X (2 - 1)) + 2) unformatted

$V = 199 - ((((|(199 - 193)| - 2)/2) \times (2-1)) + 2)$

V = 195

N = (V / X) - int(V / X) unformatted

$N = (V/X) - \int (V/X)$

N = (195 / 2) - int(195 / 2) unformatted

$N = (195/2) - \int (195/2)$

N = .5 I have a binary 1.

Last one changing R to 191

V = 195, R = 191, V - R = 4, N = .5 or binary 1

V = V + or - ((((ABS(V - R) - N) / Base) X (Base - 1)) + N) unformatted

$$V = V \pm ((((|(V - R)| - N)/Base) \times (Base - 1)) + N)$$

V = 195 - ((((ABS(195 - 193) - 2) / 2) X (2 - 1)) + 2) unformatted

$$V = 195 - ((((|(195 - 193)| - 2)/2) \times (2 - 1)) + 2)$$

V = 193

N = (V / X) - int(V / X) unformatted

$$N = (V/X) - \int (V/X)$$

N = 193 / 2) - int(193 / 2) unformatted

$$N = 193/2¿ - \int (193/2)$$

N = .5 (No I wanted a binary zero. Changing R to 191

V = 195, R = 191, V - R = 4, N = .5 or binary 1

V = V + or - ((((ABS(V - R) - N) / Base) X (Base - 1)) + N) unformatted

$$V = V \pm ((((|(V - R)| - N)/Base) \times (Base - 1)) + N)$$

V = 195 - ((((ABS(195 - 191) - 2) / 2) X (2 - 1)) + 2) unformatted

$$V = 195 - ((((|(195 - 191)| - 2)/2) \times (2 - 1)) + 2)$$

V = 192

N = (V / X) - int(V / X) unformatted

$$N = (V/X) - \int (V/X)$$

N = (192 / 2) - int(192 / 2) unformatted

$$N = (192/2) - \int (192/2)$$

N = 0.

So, I have calulated my distances I need to move R around and code 0110 in BNS/Burris Numerical system.

So, to move my R and decode the distances for 0110 between V and R here I could use 4, 6, 5 decoding from V = 199. I would start out with a zero. Now I will use R = 1 and V = (R + distance) or V = (1 + 4).

Before getting my decode distances I need to calulate my coding distances. Will going back up above and looking at the equations will give me those.

The first one is 1,

The second one is 2,

The third one is 2,

So my coding distances is 1, 2, 2.

My decoding distances is 4, 6, 5.

So here I use V = 1 R = 2 V - R = 1

Figure 20

V = V + or - ((ABS(V - R) X (Base - 1)) + N) unformatted

$$V = V \pm ((|(V-R)| \times (Base-1)) + N)$$

V = 2 + ((ABS(2 - 1) X (2 - 1)) + 2) unformatted

$$V = 2 + ((|(2-1)| \times (2-1)) + 2)$$

V = 5 R = 1 V - R = 4

R = R + 2

V = 5 R = 3 V - R = 2

V = V + or - ((ABS(V - R) X (Base - 1)) + N) unformatted

$$V = V \pm ((|(V-R)| \times (Base-1)) + N)$$

V = 5 + ((ABS(5 - 3) X (2 - 1)) + 2) unformatted

$$V = 5 + ((|(5-3)|\ X\ (2-1)) + 2)$$

V = 9, R = 3 V - R = 6

R = R + 4

V = 9, R = 7 V - R = 2

V = V + or - ((ABS(V - R) X (Base - 1)) + N) unformatted

$$V = V \pm ((|(V - R)| \times (Base - 1)) + N)$$

V = 9 + ((ABS(9 - 7) X (2 - 1)) + 1) unformatted

$$V = 9 + ((|(9 - 7)| \times (2 - 1)) + 1)$$

V = 12, R = 7, V - R = 5

So, as you see done correctly coding or decoding in BNS/Burris numerical system can be planned going forward or backwards. So basicly again to make BNS work everything becomes a matter of a distance problem to be solved between the variables V and R. If I can solve those distances in such a way so that I can always know and keep track of my distances between my variables V and R I can code a lot of information into this system. I would calulate the distances between V - R to fit into a system that conforms to a set of rules and regulations. If my V - R conforms to such a system with rules and regulations that I know then I can code as much information into BNS/Burris Numerical system as I want.

Here I am going to pick my V's first then plot my R's using the equation:

R1 = V1 + or – ((ABS(V2 - V1) - N2) / (base - 1)) is the equation for plotting R. unformatted

$$R1 = V1 + ¿ - ((|(V2 - V1)| - N2) / (base - 1))$$ is the equation for plotting R.

is the equation for plotting R.

Lets code 0100

Figure 21

V4 = 20 for a binary 0.

V3 =   17 for a binary 1.

V2 = 14 for a binary 0.

V1 = 12 for binary 0.

R4 = V3 - (((V4 - V3) - N4) / (base - 1)) unformatted

$$R4=V3-((|(V4-V3)|-N4)/(base-1))$$

R4 = 17 - ((ABS(20 - 17) - 1) / (2 – 1)) unformatted

$$R4=17-((|(20-17)|-1)/(2-1))$$

R4 = 15

V4 = 20, R4 = 15

R3 = V2 - ((ABS(V3 - V2) - N3) / (base – 1)) unformatted

$$R3=V2-((|(V3-V2)|-N3)/(base-1))$$

R3 = 14 - ((ABS(17 - 14) - 2) / (base – 1)) unformatted

$$R3=14-((|(17-14)|-2)/(base-1))$$

R3 = 13

V3 = 17, R3 = 13

R2 = V1 + or - ((ABS(V2 - V1) - N2) / (base – 1)) unformatted

$$R2=V1\pm((|(V2-V1)|-N2)/(base-1))$$

R2 = 12 - ((ABS(14 - 12) - 1) / (base – 1)) unformatted

$$R2=12-((|(14-12)|-1)/(base-1))$$

R2 = 11

V2 = 14 R2 = 11

V1 = 12 R2 = 11 (I am done plotting R's. V1 will be last result)

So now I have my R's So now l will decode V.

Figure 22

V4 = 20, R4 = 17

V3 = V4 + or - ((((ABS(V4 - R4) - N4) / Base) X (Base - 1)) + N4) unformatted

$$V3 = V4 \pm ((((|(V4 - R4)| - N4) / Base) \times (Base - 1)) + N4)$$

V3 = 20 - ((((ABS(20 - 15) - 1) / 2) X (2 - 1)) + 1) unformatted

$$V3 = 20 - ((((|(20 - 15)| - 1) / 2) \times (2 - 1)) + 1)$$

V3 = 17, R3 = 13 (I changed my R from 17 to 13)

V2 = V3 + or - ((((ABS(V3 - R3) - N3) / Base) X (Base - 1)) + N3) unformatted

$$V2 = V3 \pm ((((|(V3 - R3)| - N3) / Base) \times (Base - 1)) + N3)$$

V2 = 17 - ((((ABS(17 - 13) - 2) / 2) X (2 - 1)) + 2) unformatted

$$V2 = 17 - ((((|(17 - 13)| - 2) / 2) \times (2 - 1)) + 2)$$

V2 = 14, R2 = 11 (I changed my R from 13 to 11)

V2 = V3 + or - ((((ABS(V3 - R3) - N3) / Base) X (Base - 1)) + N3) unformatted

$$V2 = V3 \pm ((((|(V3 - R3)| - N3) / Base) \times (Base - 1)) + N3)$$

V1 = V2 + or - ((((ABS(V2 - R2) - N2) / 2) X (2 - 1)) + N2) unformatted

$$V1 = V2 \pm ((((|(V2 - R2)| - N2) / 2) \times (2 - 1)) + N2)$$

V1 = 14 - ((((ABS(14 - 11) - 1) / 2) X (2 - 1)) + 1) unformatted

$$V1 = 14 - ((((|(14 - 11)| - 1) / 2) \times (2 - 1)) + 1)$$

V1 = 12, R2 = 11

Decode Done. So, I chose my V's manually and then plotted my R's. So, some control over this is possible.

Now, when plotting R's some new things come into play.

Here is a chart with R as 1.

Figure 23

1 = 2 or 3

2 = 4 or 5

3 = 6 or 7

All the even numbers here are binary 0's and all the odd numbers here are binary 1's.

Here R = 2. So this changes the chart.

Figure 24

2 = 3 or 4

3 = 5 or 6

4 = 7 or 8

All the even numbers here are binary 1's and all the even numbers are binary 1's.

So, now if I start mixing my R's up in plotting my codes and decodes I will have to look at both V and R at the same time to decided what N is. But, by mixing it up I have even more control now over trying to create a system to control what R is as I decode.

Now, since I have shown how to manually choose V and plot R and briefly talked about mixing it up So that decoding becomes more advance I want to say one more thing. Now, to create a system of control it should be possible now to manually pick V and plot R so that R can be fit into some algorithm or decodeing chart to create a decrypting system to so that it is now possible to code and decode data making this a valuable system.

3.4Using a key dynamic or static for code and decode of BNS.

With BNS a key can be generated for coding or decoding. That key can by static meaning that it never changes or the key can be dynamic meaning a different key is generated for each code or decode. First to apply a key a distance chart may be generated. When V and R is coded the distance between V minus R is recorded. Then an algorithm is ran to adjust the distance between V minus R so that R is closer to V. The general version of this is that the distance between V and R increases times the base so when adjusting R the distance between V and R may decreased by dividing the distance into the base. So a distance key is generated showing that if the V minus R is this much then V minus R will be this much. Now, that distance chart can be taken and applied to each V going up the chart. Giving each V multiple R's before V minus R is adjusted and giving each V a multiple new R after V minus R is adjusted.

What R is given to each V after a code depends on using a distance chart to calulate what new R's will be given to V when R is adjusted during coding. This V and R chart or distance chart will be used to decode R during decode so that all the previous V's thus the data can be found.

Figure

distance chart 13,7

Applying the chart V and R to create a V and R chart.

Before After

V = 100, R = 87, V = 100, R = 93

V = 102, R = 89, V = 102, R = 95

Distance Chart 25,13

Applying the chart to V and R to create a V and R chart.

V = 100, R = 75, V = 100, R = 87

V = 102, R = 77, V = 101, R = 89

So, using a distance chart of V after coding before R is adjusted I can create a key so the V and R is found on the key and the new V and R is given so that a decode is possible. With a distance chart the key can be generated as V and R is coded.

4. RESULTS

4.1 EINSTEIN'S TRAIN THOUGHT EXPERIMENT

First lets define R in my equations as the reference point.

And lets define V in my equation as another object.

The reference point is:

Start quote:http://en.wikipedia.org/wiki/Reference

Reference is a relation between objects in which one object designates, or acts as a means by which to connect to or link to, another object. The first object in this relation is said to refer to the second object. The second object – the one to which the first object refers – is called the referent of the first object.

End of quote

Start quote:http://en.wikipedia.org/wiki/Relativity_of_simultaneity

Whether two events occur at the same time is not absolute, but depends on the observer's reference frame.

End quote:

This was what Albert Einstein showed in his famous train thought experiment. It was Albert Einstein's train thought experiment that provided me with the final piece of information I needed to make my numerical system work. Here is why and how Einstein helped me.

Each frame of reference for the observer is broken down here.

Figure 25

Frame of reference 0

Event = 0

Time = 0

Data (my event carring information. This is N or the numerical value that his a limit in its specific base.)Reference point (my observer).

Event 0 causes event 1

Frame of reference 1

Event = 1

Time = 1

Data (my event carring information. This is N or the numerical value that has a limit in its specific base.)Reference point (my observer),

Event 1 causes event 2

Frame of reference 2

Event = 2

Time = 2

V = some number event

R = some value the observer.

Event 2 causes event 3

Frame of reference 3

Event = 3

Time = 3

V = some number event.

R = some value the observer.

Now, because my events are not absolute which would give too many solutions to make my system workable, by breaking down all events into the frame of a single observer only my system can keep track of a limited amount of solutions equal to the amount of events to tracked.

Now because my events are dependent upon the observer as long as all the observers are known the events as they happened can be rewinded or played forward like a piece of video tape.

The events happening at t = 0, t = 1, t = 2 t = 3 and so forth form a timeline.

Start quote(http://en.wikipedia.org/wiki/Timeline)

A timeline is a way of displaying a list of events in chronological order,

End quote

To put it more bluntly the timeline of the events from the perspective of the reference points can be reconstructed. Now, the timeline as it played out started with the last observer and the last event can always be reconstructed on command when ever needed. Thus, information can be stored or numbers can be expressed and because time is being used here the limit of the amount of information that can be stored or the limit of the numbers that can be expressed is only limited to the amount of available time to complete the process. Space is no longer the constraint factor. Time instead of space has become the constraint factor.

The final contribution from Einstein here is:

Start Quote(http://en.wikipedia.org/wiki/Relativity_of_simultaneity)

"If the two events are causally connected ("event A causes event B"), then the relativity of simultaneity preserves the causal order (i.e. "event A causes event B" in all frames of reference)."

End Quote

Then event B causes event C and in all frames of referece in this connection between events is preserved. The chain of events as they happened are preserved. So in doing so the numbers, data, and information are preserved through the timeline of the events. The timeline containing the information is preserved and can be decoded back out as long as all the observers are known and retrieved in order of the caused events. All because the relativity of simultaneity preserves the causal order of events in all frames of reference.

If BNS is nothing else it is a basic mathematical model for Einstein's train thought experiment in action. It is also a model for how information can be stored in time and space or retrieved from time and space.

The observer at the train tracks is one reference point. The train is the event data being observed. Even after the train passes as the observer moves around in 3d space and time that is akin to the reference point moving around. What the observer sees as the observer moves around again is the event data from different perspectives which changes with time. Though in a 3d world the observer would have more than one event to look at but the concept is still the same for BNS.

Results

4.2 Defining BNS as a Number System.

Number systems are based on the max value of single digit within the system. In base ten the highest number is 9. In base two the highest number is 2. What I have presented here is a numerical system governed by equations that can use any base in a number system. I call BNS a number system as well as a numerical system. What seperates BNS from other traditional number systems is that it is designed to represent numbers in a non-traditional manner. It is designed to represent numbers in space like traditional number systems but it is also designed to represent numbers across time using concepts like the reference point observing its event.

Now while representing numbers across time the entire number can not be seen nor known until it is decoded according to its decode plan. BNS qualifies as a number system because it is a system for expressing numbers though that system again is non-traditional.

I have not shown using BNS with decimals. Largly I have decided that people will develop their own preferences for using decimals in BNS. A decimal could be used in BNS like this V42345 R42335.V37819 R37800 where the decimal is outside of the decoding process or numbers with decimals could be coded into BNS using ASCII as is done in binary on computers. Either way the algebraic and arithmetic structure of numbers can be perserved and expressed in BNS. Thus also qualifying it as a numerical system and a number system.

I could go into a long explanation of number systems the world has a long and rich history of them though many or most of them are not known or taught very much any more. I will say that BNS/Burris numerical system is designed to express numbers both as a numerical system and as a number system for the specific purpose of expressing those numbers in space as well as time.

Because of this property it may find its place as a way to store information more than a way to work with information though it is my hope in the future it will find its way equally for both uses.

4. RESULTS

4.3 BNS STORES INFORMATION IS TIME AS WELL AS SPACE.

Number systems are based on the max value of single digit within the system. In base ten the highest number is 9. In base two the highest number is 2. What I have presented here is a numerical system governed by equations that can use any base in number system. I call BNS a number system as well as a numerical system. What seperates BNS from other traditional number systems is that it is designed to represent numbers in a non-traditional manner. It is designed to represent numbers in space like traditional number systems but it is also designed to represent numbers across time using concepts like the reference point.

Now while representing numbers across time the entire number can not be seen nor known until it is decoded according to its decode plan. BNS qualifies as a number system because it is a system for expressing numbers though that system again is non-traditional.

I have not shown using BNS with decimals. Largly I have decided that people will develop their own preferences for using decimals in BNS. A decimal could be used in BNS like this V42345 R42335.V37819 R37800 where the decimal is outside of the decoding process or numbers with decimals could be coded into BNS using ASCII as is done in binary on computers. Either way the algebraic and arithmetic structure of numbers can be perserved and expressed in BNS. Thus also qualifying it as a numerical system and a number system.

I could go into a long explanation of number systems the world has a long and rich history of them though many or most of them are not known or taught very much any more. I will say that BNS/Burris numerical system is designed to express numbers both as a numerical system and as a number system for the specific purpose of expressing those numbers in space as well as time.

Because of this property it may find its place as a way to store information more than a way to work with information though it is my hope in the future it will find its way equally with both uses.

Results 4.4 Defining BNS numbers as space-time pathways.

A space-time pathway is basicly a time-line of events with their reference points in space-time. Now, I call it a space-time pathway because it is possible to have different time-lines at that specific place in space and time. But the resulting time-line is just one pathway of events in time what was taken. Again, at that specific place in space and time. So, that is what BNS does. Since it codes information as a time-line each specfic BNS number is a specific space-time pathway of events in space and time. So I call a code and decode coding or decoding a space-time pathway.

Results 4.5 All information exist in space-time.

From with work with BNS I believe that all information always exist in space and time. It is just waiting for space-time to represent that information as energy and matter. So space and time is like water and information is like the container that holds that water. Space and time always seeks to try to represent all information possible with energy and matter. So, space-time pathways exist. If it is not currently represented with energy and matter and some point in time space and time will find a way to make that information real. So, what happened in the past can be what will happen in the future and the other way around. As Univeres come and go sooner or later all information finds a way to exist as energy and matter.

4.6 Matter and Energy causes this information to materialize in reality.

When one can take an event with a reference point and decode that event by recalling past reference points to retrieve information theortically and mathematically from time then it is a wonder if all information exist in time even though it is not being represented by matter and energy. If so then it seems that space and time seeks to represent information with matter and energy. And, when information is no longer being represented with matter and energy later in space and time that information can exist again. It is almost like star treks transporter. Taking information that is represent with matter and energy and taking it apart only to put it back together again at some other place in space and time. Space-time is natures star trek transporter. Transporting information from one part of space time demateriallizing it and putting it back to gether again at another part of space time.

4.7 BNS as a causality model

BNS is a model for causality because with V and R it represent the first event the cause and the second event the effect. The second event then becomes for thirst event the cause and another event the effect. Then again the effect becomes the cause for the effect again. It is by knowning the reference points that the chain of the cause and effect events can be recalled from the end to the very beginning. BNS in representing causality in terms that Einstien proposed which allows BNS stores information and express a number mathematically across time. Only the effect with its reference point is represented with matter and energy at the end of this causality model with the rest of the information again being matematically stored in time.

4.8 BNS for encryption

Though it was not my intention to invent an encryption system BNS could be used very well in this regard. Decrypting would not be very easy because with only the last value in the number being represented there is no way to verify a correct decode without having more information. Also, in the process of cracking the encryption it is possible to get files that are perfectly good files but are not the orginal file. With BNS it is critcial to have the same method for decode that was used for encode.

4.9Misc terms.

In BNS the file size could rightly so be referred to the age of the file because it is not only the size of the file but a indicator of how long it would take to decode the file. So, since information is mathematically being store in time the file size is also refereced to as the age of the file because the file size is the time it took to code it and takes to decode it. BNS was invented from the concept of the Vector Coordinate Numerical system and should be called the Burris Numerical System or instead of VCNS BNS (Burris Numerical system).

CONCLUSION

BNS/Burris Numerical System is not a easy system for humans to use though it could be done by scribes and mathematicians on scrolls and journals as in the old days. It could be used to work with incredibly large numbers mathematically. It could be used to store hugh amounts of data. It could be used as a doomsday way to store all of mankind's knowledge in the event of a natural disaster. Or it could just be used in normal everyday computing because it is just a plain easy way to keep a lot of information in a small amount of space. And, it could be used to express numbers and perform mathimatical calulations as other numerical systems do.

Number systems/numerical system are not just for primates with fingers. Computers are great with algorithms and equations so number systems with special properties that are relative for specific situations can depend on algorithms and equations to function in their own right as a numerical system. And deserve the label numerical system or number system. BNS falls into the catagory of numerical system because it is designed for numbers of all bases and be used to work math with. Only with BNS the last part of the number is stored in space the rest of the number is mathematically stored across time using Einsteins work.

It is time to expand the concepts of number systems and numerical systems so that math as a field can continue to grow and mature as man learns more about her/his environment and develops new sciences and technologies.

Einstein's concept of reference and its relationship with its external variable the event that holds true as one event causes another event creating a chronological order of events that become a timeline in space is the model that BNS/Burris Numerical System simulates. It simulates this for the purpose of being able to back up the timeline traveling to the past mathematically (rewinding the timeline to observe the events) or go forward in time to observe new events with the ability to rewind at any time. And in the end all BNS/Burris Numerical system needs to retrieve the action between the events and its reference point is the very last event and the very last reference point. Why, because the relationship between the reference points and the events hold true in every frame of causality. In every frame of reference.

What the Burris Numerical System does is takes information as the events and codes it with a references point turning it into a space-time pathway with only the past place value of this number being physically created with matter and energy. The rest of the number is mathematically stored across time because the last frame of reference holds true to the very first frame of reference by its reference points. So it is a structure existing in time Now, as the space-time pathway is decoded the events that happened in that time-line are written to the computers hard drive as a computer file.

And a computer file when coded into a BNS number is converted from a computer file to a space-time pathway.

Literature Cited

http://en.wikipedia.org

Appendices

Programs I used to research and develop the Burris Numerical system. Please excuse some of my statements I did not know I would be publishing my research at the time. I edited for some of the bad words and took them out. Note: I was limited for publication to 630 pages so I was not able to put in all my research. So I have 978 pages left out. I will try to publish the left out research sometime in the future.

```
"C:\Users\Reactor1967\vcns\work\R060103.BAS"

CLS

t$ = "c:\db.txt"

OPEN t$ FOR INPUT AS #1

DO

INPUT #1, lb#, realdeal$, crank#

PRINT lb#

PRINT realdeal$

PRINT crank#
```

```
a$ = INKEY$

REM INPUT a$

IF a$ = "s" THEN STOP

LOOP UNTIL z2 = -1

CLOSE
```

"C:\Users\Reactor1967\vcns\work\PROOF.BAS"

```
REM this program proves that you can code to what ever distance you want

REM ending dist# = beginning (dist# * 2) + N

v# = 1273

r# = 574

CLS

RANDOMIZE TIMER

PRINT (v# - r#)

z = INT(RND * 2) + 0

v# = v# + (v# - r#) + z + 1

PRINT (v# - r#)

r# = v# - 5: REM 5 is beginning distance. Out come will be 11 or 12 always

z = INT(RND * 2) + 0

v# = v# + (v# - r#) + z + 1

PRINT v# - r#
```

"C:\Users\Reactor1967\vcns\work\O72203A.BAS"

```
REM first program decoding according to 7-20-03 method. Cool.
```

```
RANDOMIZE TIMER

v# = 1

r# = 1

CLS

DO

REM z = INT(RND * 2) + 0

REM IF z = 0 THEN v# = v# + (v# - r# + 0)

REM IF z = 1 THEN v# = v# + (v# - r# + 2)

v# = v# + (v# - r# + 2)

PRINT , , v#; r#; v# - r#; d#

a$ = INKEY$

REM INPUT a$

IF a$ = "s" THEN STOP

IF a$ = "d" THEN EXIT DO

d# = (v# - r#) * .5

r# = v# - d#

LOOP

DO

PRINT , , v#; r#; v# - r#; d#

IF v# = 1 THEN EXIT DO

a$ = INKEY$

REM INPUT a$

IF a$ = "s" THEN STOP

d# = (v# - r#)
```

```
IF (d# / 2) - INT(d# / 2) THEN d# = d# - 1
d# = d# / 2
d# = d# + 1
v# = v# - d#
d# = d# - 2
d# = (d# / .5)
r# = v# - d#
LOOP
```

```
"C:\Users\Reactor1967\vcns\work\MILEMARK.BAS"
v# = 1
r# = 1
CLS
RANDOMIZE TIMER
count = 0
PRINT , , v#; r#; v# - r#
dist# = 0
DO
count = count + 1
z = INT(RND * 2) + 0
IF z = 0 THEN dist# = dist# + 1
IF z = 1 THEN dist# = dist# + 2
IF z = 0 THEN v# = v# + (v# - r# + 2)
IF z = 1 THEN v# = v# + (v# - r# + 4)
```

```
PRINT z; v#; r#; v# - r#; v# - sv#; r# - sr#; dist#

decode# = (r# - sr#)

sv# = v#

sr# = r#

a$ = INKEY$

REM INPUT a$

IF a$ = "s" THEN STOP

r# = r# + dist#

LOOP
```

```
"C:\Users\Reactor1967\vcns\work\MAP.BAS"

REM This demonstrates the map ideal in the book vcns.

REM this program has a lower starting limit as long as you know your

REM limits you can aways find the value of n

REM when you increase r increase it so that your map stays inline.

v# = 2

r# = 1

z = 0

CLS

RANDOMIZE TIMER

DO

PRINT "Demo of Map ideal lower limit for v# = 2"
```

```
IF z = 0 THEN z = 1 ELSE z = 0

z = INT(RND * 2) + 0

PRINT , , z; v#; v# + (v# - r# + 1); v# + (v# - r# + 3)

INPUT a$

IF a$ = "s" THEN STOP

v# = v# + 2

LOOP
```

```
"C:\Users\Reactor1967\vcns\work\MANUALC.BAS"

v# = 1

r# = 1

spr# = 0

RANDOMIZE TIMER

CLS

DO

z = INT(RND * 2) + 0

v# = v# + (v# - r#) + z + 1

PRINT v#; r#; v# - r#; (v# - r#) - spr2#; spr#

spr2# = (v# - r#)

a$ = INKEY$

REM INPUT a$

IF a$ = "s" THEN STOP
```

```
d1# = (v# - r#) - (spr# + 0)

d2# = (v# - r#) - (spr# + 2)

d3# = (v# - r#) - (spr# + 4)

flag = 0

DO

IF d3# >= (spr# + 4) THEN flag = 3

IF flag = 3 THEN EXIT DO

IF d2# >= (spr# + 2) THEN flag = 2

IF flag = 2 THEN EXIT DO

IF d1# >= (spr# + 0) THEN flag = 1

IF flag = 1 THEN EXIT DO

STOP

LOOP

IF flag = 3 THEN spr# = spr# + 4

IF flag = 2 THEN spr# = spr# + 2

IF flag = 1 THEN spr# = spr# + 0

r# = r# + spr#

LOOP
```

```
"C:\Users\Reactor1967\vcns\work\LLOYD.BAS"

v# = 1

r# = 1
```

```
CLS
RANDOMIZE TIMER
DO
z = INT(RND * 2) + 0
test = v# + (v# - r#) + 1 >= (r# + 100)
IF test = -1 THEN z = 0
IF flag = 1 THEN z = 2
IF z = 0 THEN v# = v# + (v# - r# + 1)
IF z = 1 THEN v# = v# + (v# - r# + 2)
IF z = 2 THEN v# = v# + (v# - r# + 3)
PRINT , , z; v#; r#; v# - r#
IF v# = 1 THEN STOP
IF r# > v# THEN STOP
a$ = INKEY$
REM INPUT a$
IF a$ = "s" THEN STOP
IF test = -1 THEN v# = v# - 101
IF test = -1 THEN flag = 1 ELSE flag = 0
LOOP
```

```
"C:\Users\Reactor1967\vcns\work\LBSPEC.BAS"
v# = 1
```

```
r# = 1

CLS

RANDOMIZE TIMER

DO

z = INT(RND * 2) + 0

dist# = v# - r#

IF z = 0 THEN v# = v# + (v# - r# + 1)

IF z = 1 THEN v# = v# + (v# - r# + 2)

x# = dist#

PRINT z; v#; r#; dist#; r# - sr#; v# - r#; v# - sv#

sv# = v#

sr# = r#

r# = r# + dist#

a$ = INKEY$

IF a$ = "s" THEN STOP

LOOP
```

```
"C:\Users\Reactor1967\vcns\work\LADDER.BAS"

REM the distances which will represent v - r must have a zero

REM beside them to be usable. On a other note if the distances

REM have a .5 beside them those or the distances that crossed

REM over from a 0 to a 1 or a 1 to a 0.
```

```
d# = 2

CLS

status = 1

DO

d# = d# + 1

PRINT , , d#; ((d# - 3) / 4) - INT((d# - 3) / 4)

IF status = 1 THEN PRINT , , , "0 from a 0"

IF status = 2 THEN PRINT , , , "1 from a 0"

IF status = 3 THEN PRINT , , , "0 from a 1"

IF status = 4 THEN PRINT , , , "1 from a 1"

IF status = 4 THEN status = 0

a$ = INKEY$

REM INPUT a$

status = status + 1

IF a$ = "s" THEN STOP

d# = d# + 1

PRINT , , d#; ((d# - 2) / 4) - INT((d# - 2) / 4)

IF status = 1 THEN PRINT , , , "0 from a 0"

IF status = 2 THEN PRINT , , , "1 from a 0"

IF status = 3 THEN PRINT , , , "0 from a 1"

IF status = 4 THEN PRINT , , , "1 from a 1"

IF status = 4 THEN status = 0

a$ = INKEY$

REM INPUT a$
```

```
status = status + 1

IF a$ = "s" THEN STOP

LOOP
```

"C:\Users\Reactor1967\vcns\work\DXCAL.BAS"

```
REM process for finding x# which is the distance you code when. you

REM can divide x# by 4 to get a fractional value(remainder) which tells

REM you how much to subtract from x# to get your speed of r#. Take
your

REM speed and subtract it from your current r# to get your previous r#

d# = 247

CLS

DO

x# = d#

x# = x# - 1

x# = x# / 2

x# = x# - 1: REM -1 can be anything.

REM         positive or negative

REM         this determines the distance between d# & x#

spr# = d# - x#

PRINT d#; x#; spr#; x# - spr#; x# / 4

INPUT a$
```

```
IF a$ = "s" THEN STOP

d# = d# + 2

LOOP




"C:\Users\Reactor1967\vcns\work\DTEST.BAS"

CLS

REM 1 distances

a# = 12016

b# = 12020

c# = 12024

d# = 12036

e# = 12040

f# = 12044

REM 0 distances

g# = 12026

h# = 12028

i# = 12030

j# = 12032

k# = 12046

l# = 12052

div# = 1

DO
```

```
t1# = (a# / div#) - INT(a# / div#)

t2# = (b# / div#) - INT(b# / div#)

t3# = (c# / div#) - INT(c# / div#)

t4# = (d# / div#) - INT(d# / div#)

t5# = (e# / div#) - INT(e# / div#)

t6# = (f# / div#) - INT(f# / div#)

s1# = (g# / div#) - INT(g# / div#)

s2# = (h# / div#) - INT(h# / div#)

s3# = (i# / div#) - INT(i# / div#)

s4# = (j# / div#) - INT(j# / div#)

s5# = (k# / div#) - INT(k# / div#)

s6# = (l# / div#) - INT(l# / div#)

test1 = (t1# = t2#) AND (t2# = t3#) AND (t3# = t4#) AND (t4# = t5#)
AND (t5# = t6#)

test2 = (s1# = s2#) AND (s2# = s3#) AND (s3# = s4#) AND (s4# = s5#)
AND (s5# = s6#)

test3 = (test1 = -1) AND (test2 = -1)

IF test3 = -1 THEN PRINT div#

REM IF test3 = -1 THEN STOP

div# = div# + 1

a$ = INKEY$

IF a$ = "s" THEN STOP

LOOP
```

"C:\Users\Reactor1967\vcns\work\DRCHART.BAS"

```
REM coding and incoding chart

REM c1# - i1# is what r# increases by for the dist# c1

REM c2# - i2# is what r# increasey by for the dist c2#

REM when decoding if your distance is o3# or o4# r# = r# - (i2# - c2#)

REM when decoding if your distance is o1# or o2# then r# = r# - (i1# - c1#)

c# = 2

CLS

DO

i1# = (c# / 2)

o1# = ((c# - i1#) * 2) + 1

o2# = ((c# - i1#) * 2) + 2

c1# = c#

c# = c# + 1

c2# = c#

i2# = (c# / 2) - .5

o3# = ((c# - i2#) * 2) + 1

o4# = ((c# - i2#) * 2) + 2

PRINT c1#; c2#; ; c1# - i1#; c2# - i2#; ; o1#; o2#; ; o3#; o4#

a$ = INKEY$

REM INPUT a$

IF a$ = "s" THEN STOP
```

c# = c# + 1

LOOP

"C:\Users\Reactor1967\vcns\work\distrfind.txt"

1. before you code find your distance you are coding to
2. the dist# to use from v# to find r# before coding.
3. the speed of r#
4. divide distance coding to tell how much to subtract from
   dist# to find speed of r#

237 / 3 = 0 3 difference (dist# - (((dist# - 1) / 2) + 2))

120 117

239 / 3 = .666 1 difference (dist# - (((dist# - 1) / 2) + 1))

120 119

240 / 2 = even 0 difference (add# nothing)

120 120

242 / 2 = odd 2 difference (add# 1 to left)

122 120

"C:\Users\Reactor1967\vcns\work\dist.txt"

2,3,4,5 = 2

6,7,8,9 = 4

10,11,12,13 = 6

speed of r# = dist#

2 = 2,3,4,5

4 = 6,7,8,9

6 = 10,11,12,13

"C:\Users\Reactor1967\vcns\work\DESTIN.BAS"

```
t$ = "e:\lb.txt"

CLS

RANDOMIZE TIMER

x1 = 0

x2 = 0

x3 = 200

v# = 1

r# = 1

redo:

DO

z = INT(RND * 2) + 0
```

```
CLOSE #1

OPEN t$ FOR INPUT AS #1

FOR count = 1 TO 400

INPUT #1, z1, z2, z3, z4, z5, z6

IF (z4 / 2) - INT(z4 / 2) = .5 THEN flag = 1

IF (z4 / 2) - INT(z4 / 2) = 0 THEN flag = 0

test1 = (z1 = x1) AND (z2 = x2) AND (z3 = x3) AND (z = 0) AND (flag
= 0)

test2 = (z1 = x1) AND (z2 = x2) AND (z3 = x3) AND (z = 1) AND (flag
= 1)

IF test1 = -1 THEN EXIT FOR

IF test2 = -1 THEN EXIT FOR

a$ = INKEY$

REM INPUT a$

IF a$ = "s" THEN STOP

NEXT count

x1 = z4

x2 = z5

x3 = z6

STOP

IF z = 0 THEN v# = v# + (v# - r# + 1)

IF z = 1 THEN v# = v# + (v# - r# + 2)

test = (v# >= (r# + 99))

IF test = -1 THEN r# = r# + 100

PRINT z; v#; r#; v# - r#; "= "; z1; z2; z3; x1; x2; x3; flag
```

```
a$ = INKEY$
REM INPUT a$
IF a$ = "s" THEN STOP
IF a$ = "d" THEN EXIT DO

LOOP
```

```
"C:\Users\Reactor1967\vcns\work\DELETEME.BAS"
CLS
t$ = "c:\db.txt"
REM INPUT #1, lb#, realdeal$, crank#
v# = 1
r# = 1
RANDOMIZE TIMER
CLS
bin$ = ""
redo:
start = (v# - r#)
DO
z = INT(RND * 2) + 0
bin$ = bin$ + STR$(z)
IF z = 0 THEN v# = v# + (v# - r# + 1)
```

```
IF z = 1 THEN v# = v# + (v# - r# + 2)

PRINT , z; v#; r#; v# - r#

INPUT a$

IF a$ = "s" THEN STOP

test = v# >= (r# + 99)

IF test = -1 THEN EXIT DO

LOOP

OPEN t$ FOR INPUT AS #1

DO

INPUT #1, z1, g$, z2

PRINT g$

a$ = INKEY$

REM INPUT a$

IF a$ = "s" THEN STOP

test = (g$ = bin$)

IF z1 = -1 THEN STOP

IF g$ = "-1" THEN STOP

IF z2 = -1 THEN STOP

IF test = -1 THEN EXIT DO

LOOP

CLOSE

PRINT bin$, g$

IF test = -1 THEN PRINT "YOU MADE IT DUDE!"
```

```
"C:\Users\Reactor1967\vcns\work\DDTBASE.BAS"
z = -1
d# = 2
t$ = "c:\test.txt"
CLS
DO
a# = INT(d# / 2)
REM a# = d#
OPEN t$ FOR APPEND AS #1
d1# = 1
d2# = 2
WRITE #1, d#
FOR count = 1 TO a#
WRITE #1, (d1# / d#) - INT(d1# / d#), (d2# / d#) - INT(d2# / d#)
a$ = INKEY$
IF a$ = "s" THEN CLOSE #1
IF a$ = "s" THEN SYSTEM
d1# = d1# + 1
d2# = d2# + 2
NEXT count
WRITE #1, z
```

```
CLOSE #1

PRINT "Written d = "; d#

a$ = INKEY$

REM INPUT a$

IF a$ = "s" THEN CLOSE #1

IF a$ = "s" THEN SYSTEM

d# = d# + 1

IF d# = 101 THEN EXIT DO

LOOP
```

```
"C:\Users\Reactor1967\vcns\work\DBSEARCH.BAS"

t$ = "c:\db.txt"

redo:

CLS

INPUT t1

INPUT t2

OPEN t$ FOR INPUT AS #1

flag = 0

DO

INPUT #1, z1, z2, a$, z3, z4

test = (t1 = z3) AND (t2 = z4)

IF test = -1 THEN flag = 1
```

```
IF test = -1 THEN PRINT z1; z2; a$; z3; z4

IF z1 = 999 THEN EXIT DO

a$ = INKEY$

IF a$ = "s" THEN STOP

LOOP

CLOSE #1

IF flag = 0 THEN PRINT "No match"

PRINT "Again, y/n"

INPUT a$

IF a$ = "y" THEN GOTO redo:

STOP
```

```
"C:\Users\Reactor1967\vcns\work\DB.TXT"

86

158

87

161

95

185

265

325
```

81

144

50

143

97

190

277

350

94

181

256

98

195

288

374

446

90

170

230

76

128

73

93

178

249

82

147

59

72

89

168

226

63

51

88

163

214

169

227

66

65

60

78

135

53

164

217

167

224

56

180

254

96

188

271

337

62

194

287

372

443

83

150

67

69

75

100

200

299

397

494

588

675

750

99

197

293

385

468

535

228

71

84

152

326

213

223

286

370

437

126

215

61

92

175

241

166

221

222

80

140

149

17893,17701

"C:\Users\Reactor1967\vcns\work\cntadd.txt"

0 1

0 + 2

0 1

1 + 1

1 2

0 + 1

1 2

1 + 2

"C:\Users\Reactor1967\vcns\work\CDCHART1.BAS"

REM !!!!!!!!!!!!!!!!!!!!!!!!!!!!IMPORTANT BREAKTHROUGH!!!!!!!!!!!!!!!!!!!!!

REM I,ve don this program before a long time ago and here I am again back

REM at it but this time I know something different. Fist if you know the

REM distance your going to code before hand you can control where v and r

REM get closer or farther apart. I knew that before. Here is what I learn

REM that is new. A few weeks ago I realized that the distance v is from

REM r can be used as an indicator to tell the n value of v. With this

REM in mind I can use the distance to tell N and work on knowing what

REM Im coding to depending if im going from a 0 to a 0 or a 0 to a 1

REM or a 1 to a 0 or a 1 to a 0. N will only have two choices to code

REM to but distances will have base^2 squared choices.

REM now here is how all this helps me. When I decode a vector or v#

REM I look at v# - r# before I decode and that tells me the N value

REM of v#. I then decode V# and look at the v# - r# again with the

REM same R# and that tells me again what N is for the value v# is

REM currently at. So I know N before hand and I know N after the fact.

REM Now I can take those two N values and know what my previous v# - r#

REM should be. This program is as close as I can get right now to

REM a prototype of that type of decode system.

```
REM use this chart for the next program as a tool for coding and
decoding.

REM ------------------------------------------------------------------

REM rules if v - r = odd then n = 1

REM rules if v - r = even then n = 2

REM rules binary 0 coded to binary 0 (v - r) = (v - r) + 4

REM rules binary 1 coded to binary 1 (v - r) = (v - r) + 4

REM rules binary 0 coded to binary 1 (v - r) = (v - r) + 5

REM rules binary 1 coded to binary 0 (v - r) = (v - r) + 5

REM rules previous (v - r) = ((even v - r) / 2) - 1

REM rules previous (v - r) = ((odd v - r) - 1) / 2

REM -------------------------------------------------------------------

t1 = 0

t2 = 1

d1# = 1

d2# = 2

v# = 1000

CLS

DO

r1# = v# - d1#

r2# = v# - d2#

PRINT , , t1; v# - r1#; (v# + (v# - r1# + 1)) - r1#; (v# + (v# - r1# + 2)) -
r1#; "|"; d1#

PRINT , , t2; v# - r2#; (v# + (v# - r2# + 1)) - r2#; (v# + (v# - r2# + 2)) -
r2#; "|"; d2#
```

```
d1# = d1# + 2
d2# = d2# + 2
a$ = INKEY$
INPUT a$
IF a$ = "s" THEN STOP
LOOP
```

```
"C:\Users\Reactor1967\vcns\work\BINSUB.BAS"
CLS
REM INPUT "Please enter count"; count
count = 1
DO
d# = 1
count2 = 1
DO
IF d# >= count THEN EXIT DO
d# = d# * 2
count2 = count2 + 1
a$ = INKEY$
IF a$ = "s" THEN STOP
LOOP
IF d# > count THEN count2 = count2 - 1
```

```
store = count
count2 = count2 - 1
bin$ = ""
DO
a# = (2 ^ count2)
REM IF a# <= count THEN PRINT "1";
IF a# <= count THEN bin$ = bin$ + "1"
REM IF a# > count THEN PRINT "0";
IF a# > count THEN bin$ = bin$ + "0"
IF a# <= count THEN count = count - a#
count2 = count2 - 1
IF count2 < 0 THEN EXIT DO
a$ = INKEY$
IF a$ = "s" THEN STOP
LOOP
PRINT , , ; store; bin$
count = store + 1
a$ = INKEY$
IF a$ = "s" THEN STOP
LOOP
```

"C:\Users\Reactor1967\vcns\work\BASE1000.BAS"

```
v# = 1
r# = 1
RANDOMIZE TIMER
CLS
DO
z = INT(RND * 1000) + 0
test = (v# - r#) >= 2
r2# = r# + ((v# - r#) * 999)
v# = v# + ((v# - r#) * 999) + z + 1
PRINT z; v#; r#; v# - r#
IF test = -1 THEN r# = r2#
a$ = INKEY$
IF a$ = "s" THEN STOP
LOOP
```

"C:\Users\Reactor1967\vcns\work\BASE100.BAS"

```
v# = 1
r# = 1
CLS
RANDOMIZE TIMER
DO
z = INT(RND * 100) + 0
```

```
test = (v# - r#) >= 2

r2# = r# + ((v# - r#) * 99)

v# = v# + ((v# - r#) * 99) + z + 1

REM PRINT z; v#; r#; v# - r#; v# - sv#; r# - sr#; (r# - sr#) - x#

PRINT z; v#; r#; v# - r#

x# = (r# - sr#)

sv# = v#

sr# = r#

IF test = -1 THEN r# = r2#

a$ = INKEY$

IF a$ = "s" THEN STOP

LOOP
```

```
"C:\Users\Reactor1967\vcns\work\BASE10.BAS"

v# = 1

r# = 1

sv# = 1

sr# = 1

CLS

RANDOMIZE TIMER

DO

z = INT(RND * 10) + 0
```

```
test = (v# - r#) >= 100

r2# = r# + ((v# - r#) * 9)

dist# = (v# - r#)

REM you and also just simply do v# = v# + ((v# - r#) * base - 1) + z + 1

IF z = 0 THEN v# = v# + ((v# - r#) * 9) + 1

IF z = 1 THEN v# = v# + ((v# - r#) * 9) + 2

IF z = 2 THEN v# = v# + ((v# - r#) * 9) + 3

IF z = 3 THEN v# = v# + ((v# - r#) * 9) + 4

IF z = 4 THEN v# = v# + ((v# - r#) * 9) + 5

IF z = 5 THEN v# = v# + ((v# - r#) * 9) + 6

IF z = 6 THEN v# = v# + ((v# - r#) * 9) + 7

IF z = 7 THEN v# = v# + ((v# - r#) * 9) + 8

IF z = 8 THEN v# = v# + ((v# - r#) * 9) + 9

IF z = 9 THEN v# = v# + ((v# - r#) * 9) + 10

REM PRINT z; v#; r#; v# - r#; (v# - r#) - x3#; dist#; dist# - x2#; v# - sv#;
(v# - sv#) - x4#; r# - sr#; (r# - sr#) - x#

PRINT z; v#; r#; v# - r#

x# = (r# - sr#)

x2# = (dist#)

x3# = (v# - r#)

x4# = (v# - sv#)

sv# = v#

sr# = r#

a$ = INKEY$
```

```
REM INPUT a$
IF a$ = "s" THEN STOP
IF test = -1 THEN r# = r2#
LOOP
```

"C:\Users\Reactor1967\vcns\work\BASE3.BAS"

```
REM my first sucessful attempt at getting r# to increase in
REM other bases
v# = 1
r# = 1
CLS
RANDOMIZE TIMER
DO
z = INT(RND * 3) + 0
test = (v# - r#) >= 30
IF test = -1 THEN flag = 1 ELSE flag = 0
r2# = r# + ((v# - r#) * 2)
IF z = 0 THEN v# = v# + ((v# - r#) * 2) + 1
IF z = 1 THEN v# = v# + ((v# - r#) * 2) + 2
IF z = 2 THEN v# = v# + ((v# - r#) * 2) + 3
PRINT z; v#; r#; v# - r#; v# - sv#; r# - sr#
sv# = v#
```

```
sr# = r#
IF flag = 1 THEN r# = r2#
a$ = INKEY$
IF a$ = "s" THEN STOP
LOOP
```

"C:\Users\Reactor1967\vcns\work\100703A.BAS"

```
REM I now know that the speed of v minus r is equal to the speed of d
REM times the base plus or minus the difference in N. So knowing that the
REM thing to do now is determine how to increase or decrease d so that
REM the speed of r is divisible by the same number every time to keep
REM the syemetery.
v# = 1
r# = 1
CLS
RANDOMIZE TIMER
DO
z = INT(RND * 2) + 0
REM z = 0
d# = v# - r#
IF z = 0 THEN v# = v# + (v# - r#) + 1
```

```
IF z = 1 THEN v# = v# + (v# - r#) + 2

m# = (v# / 4) - INT(v# / 4): m# = m# * 10: m# = INT(m#)

PRINT , z; m#; v#; r#; v# - r#; d#; d# - sd#; r# - sr#

sr# = r#

sd# = d#

d# = INT((v# - r#) / 2)

DO

d# = d# + 1

s# = (v# - r#) - d#

test = (s# / 4) - INT(s# / 4) = 0

IF test = -1 THEN EXIT DO

a$ = INKEY$

IF a$ = "s" THEN STOP

LOOP

r# = v# - d#

LOOP
```

"C:\Users\Reactor1967\vcns\work\100603A.BAS"

```
REM was going to play with rate change of (v# - r#) = rate change of d# +- rate of change of N.

v# = 1

d# = 0
```

```
RANDOMIZE TIMER

r# = v# - d#

z = 0

DO

z = INT(RND * 2) + 0

REM IF z = 0 THEN z = 1 ELSE z = 0

IF z = 0 THEN v# = v# + (v# - r# + 1)

IF z = 1 THEN v# = v# + (v# - r# + 2)

PRINT , , z; v#; r#; v# - r#; (v# - r#) - svr#; d#; d# - sd#; r# - sr#

sr# = r#

svr# = (v# - r#)

sd# = d#

a$ = INKEY$

INPUT a$

IF a$ = "s" THEN STOP

d# = d# + 2

REM d# = INT(RND * 100) + 1

r# = v# - d#

LOOP
```

"C:\Users\Reactor1967\vcns\work\100503D.BAS"

```
REM NEW FORMULA --- (v# - r#) = rate of change of (d#) * base +-
(rate of change of N)

v# = 1

r# = 1

CLS

count = 0

DO

count = count + 1

z = INT(RND * 2) + 0

z = 1

d# = (v# - r#)

IF z = 0 THEN v# = v# + (v# - r# + 1)

IF z = 1 THEN v# = v# + (v# - r# + 2)

PRINT z; v#; r#; v# - r#; d#; r# - sr#

IF count = 100 THEN EXIT DO

d# = (v# - r#)

sr# = r#

a$ = INKEY$: IF a$ = "s" THEN STOP

dist# = INT((v# - r#) / 2)

DO

dist# = dist# + 1

s# = d# - dist#

test = (s# / 4) - INT(s# / 4) = 0

IF test = -1 THEN EXIT DO
```

```
a$ = INKEY$: IF a$ = "s" THEN STOP

LOOP

r# = v# - dist#

LOOP

sr# = (r# - sr#)

DO

d# = d# - 8

REM sr# = sr# - 8

r# = v# - d#

IF z = 0 THEN v# = v# + (v# - r# + 1)

IF z = 1 THEN v# = v# + (v# - r# + 2)

PRINT , , z; v#; r#; v# - r#; d#; r# - sr2#

sr2# = r#

a$ = INKEY$

INPUT a$

IF a$ = "s" THEN STOP

LOOP
```

"C:\Users\Reactor1967\vcns\work\100503C.BAS"

```
REM YOU CAN BUILD ON THIS. GET THE SPEED OF R# TO BE DIVISIBLE BY 4.

v# = 1007
```

```
r# = 25
RANDOMIZE TIMER
CLS
redo:
sr# = r#
DO
dist# = v# - r#
dist# = dist# - 8
IF (dist# / 2) - INT(dist# / 2) = .5 THEN dist# = dist# - 1
dist# = dist# / 2
dist# = dist# + 2
dist# = dist# - 2
r# = v# - dist#
j# = v# - r#
v# = v# + (v# - r# + 1)
PRINT , , "0"; v#; r#; v# - r#; j#; r# - sr#
sr# = r#
REM IF (r# / 2) - INT(r# / 2) = 0 THEN STOP
IF v# - r# < 10 THEN EXIT DO
a$ = INKEY$
INPUT a$
IF a$ = "s" THEN STOP
LOOP
PRINT "Ready for data"
```

```
a$ = INKEY$

INPUT a$

IF a$ = "s" THEN STOP

sr# = r#

DO

z = INT(RND * 2) + 0

test = (v# + (v# - r# + 2)) - r# >= 1000

IF test = -1 THEN z = 1

j# = v# - r#

IF z = 0 THEN v# = v# + (v# - r# + 1)

IF z = 1 THEN v# = v# + (v# - r# + 2)

PRINT , , z; v#; r#; v# - r#; j#; r# - sr#

sr# = r#

a$ = INKEY$

INPUT a$

IF a$ = "s" THEN STOP

LOOP UNTIL v# - r# >= 1000

PRINT "Ready for control"

a$ = INKEY$

INPUT a$

IF a$ = "s" THEN STOP

GOTO redo:
```

```
"C:\Users\Reactor1967\vcns\work\100503B.BAS"

REM This program makes up a random data code and a control code.

REM If I can get this program to do both then I can get this coding data.

REM the control code is to bring v# and r# within operating distance of

REM each other. The random code is for coding data. Now I just need to

REM get the program to know when it is switching from one to the other.

v# = 3

r# = 1

CLS

RANDOMIZE TIMER

redo:

PRINT "DATA PHASE --------------------------------------------------"

count = 0

DO

count = count + 1

z = INT(RND * 2) + 0

test = ((v# + (v# - r# + 2)) - r#) >= 1000

IF test = -1 THEN z = 1

IF z = 0 THEN v# = v# + (v# - r# + 1)

IF z = 1 THEN v# = v# + (v# - r# + 2)

PRINT , , z; v#; r#; v# - r#; count

a$ = INKEY$

IF a$ = "s" THEN STOP
```

```
LOOP UNTIL v# - r# >= 1000
PRINT "CONTROL PHASE ------------------------------------------------"
z$ = INKEY$
INPUT z$
IF z$ = "s" THEN STOP
REM ----------------------
dist# = INT((v# - r#) / 2)
count = 0
DO
x# = v# - r#
lb# = dist#
count = count + 1
DO
test = (dist# * 2) + 1 < x#
IF test = -1 THEN EXIT DO
dist# = dist# - 1
a$ = INKEY$: IF a$ = "s" THEN STOP
LOOP
r# = v# - dist#
IF lb# - dist# = 2 THEN z = 1
IF lb# - dist# = 1 THEN z = 0
IF lb# - dist# = 0 THEN STOP
IF lb# - dist# > 2 THEN STOP
IF z = 0 THEN v# = v# + (v# - r# + 1)
```

```
IF z = 1 THEN v# = v# + (v# - r# + 2)

PRINT , , z; v#; r#; v# - r#; dist#; count

IF v# - r# < 3 THEN EXIT DO

a$ = INKEY$

REM INPUT a$

IF a$ = "s" THEN STOP

LOOP

REM PRINT "Ready for coding": rem -------------------------------------

z$ = INKEY$

REM INPUT z$

IF z$ = "s" THEN STOP

GOTO redo:
```

```
"C:\Users\Reactor1967\vcns\work\100503A.BAS"

v# = 100

r# = 99

d# = 0

CLS

RANDOMIZE TIMER

redo:

DO

z = INT(RND * 2) + 0
```

```
x# = v# - r#

DO

d# = d# + 1

d2# = (d# * 2) + z + 1

test = (d2# > x#)

IF test = -1 THEN EXIT DO

a$ = INKEY$

IF a$ = "s" THEN STOP

LOOP

r# = v# - d#

IF z = 0 THEN v# = v# + (v# - r# + 1)

IF z = 1 THEN v# = v# + (v# - r# + 2)

PRINT , , z; v#; r#; v# - r#; d#; x# - d#; d# - (x# - d#)

a$ = INKEY$

IF a$ = "s" THEN STOP

IF a$ = "d" THEN EXIT DO

LOOP

DO

DO

z = INT(RND * 2) + 0

x# = v# - r#

d# = d# - 1

d2# = (d# * 2) + z + 1

test = (d2# < x#)
```

```
IF test = -1 THEN EXIT DO

a$ = INKEY$

IF a$ = "s" THEN STOP

LOOP

r# = v# - d#

IF z = 0 THEN v# = v# + (v# - r# + 1)

IF z = 1 THEN v# = v# + (v# - r# + 2)

PRINT , , z; v#; r#; v# - r#; d#; x# - d#; (x# - d#) - d#

a$ = INKEY$

IF a$ = "s" THEN STOP

IF a$ = "d" THEN EXIT DO

LOOP

GOTO redo:
```

"C:\Users\Reactor1967\vcns\work\100403B.BAS"

```
REM Program works just fine until data is coded for some unknown
reason

REM the data is messing things up when there is no reason for it to

REM mess things up. This program should be able to function with data.

REM Problem is you can,t code both a zero or a one for the same

REM d# - sr# position.

DECLARE SUB decodecheck (v#, r#, d#, sr#, a1#, a2#, a3#)
```

```
v# = 1

r# = 1

sr# = 0

RANDOMIZE TIMER

CLS

cracker = 0

DO

d# = v# - r#

REM z = INT(RND * 2) + 0

IF (d# - sr#) = -1 THEN PRINT "Data after this line ------------------"

burris = (d# - sr# = -1)

IF z = 0 THEN v# = v# + (v# - r# + 1)

IF z = 1 THEN v# = v# + (v# - r# + 2)

m# = (v# / 4) - INT(v# / 4)

m# = m# * 10

m# = INT(m#)

PRINT z; m#; v#; r#; v# - r#; d#; sr#; d# - sr#; burris

REM IF cracker = 1 THEN CALL decodecheck(v#, r#, d#, sr#, a1#, a2#,
a3#)

a1# = v#: a2# = r#: a3# = sr#

a$ = INKEY$

INPUT a$

IF a$ = "s" THEN STOP

IF a$ = "d" THEN EXIT DO
```

```
r# = r# + sr#

test = r# + (sr# + 7) <= v#

lb# = sr#

IF test = -1 THEN sr# = sr# + 8

IF test = 0 THEN z = 0

IF test = -1 THEN z = 1

cracker = 1

LOOP

x# = (v# / 4) - INT(v# / 4)

x# = x# * 10

x# = INT(x#)

IF x# = 0 THEN z = 0

IF x# = 5 THEN z = 0

IF x# = 2 THEN z = 1

IF x# = 7 THEN z = 1

dist# = (v# - r#)

IF (dist# / 2) - INT(dist# / 2) = .5 THEN dist# = dist# - 1

dist# = dist# / 2

dist# = dist# + 1

IF z = 0 THEN dist# = dist# - 1

IF z = 1 THEN dist# = dist# - 2

d# = dist#

REM ---------------------------------------------------------
```

```
REM -------------------------------------------------------
IF d# - sr# = -1 THEN flag = 1 ELSE flag = 0
dist# = (v# - r#)
IF (dist# / 2) - INT(dist# / 2) = .5 THEN dist# = dist# - 1
dist# = dist# / 2
dist# = dist# + 1
v# = v# - dist#
IF flag = 1 THEN sr# = sr# - 8
IF flag = 1 THEN GOTO skip:
IF z = 1 THEN sr# = sr# - 8
skip:
IF sr# < 0 THEN sr# = 0
r# = r# - sr#
IF r# < 1 THEN r# = 1

SUB decodecheck (v#, r#, d#, sr#, a1#, a2#, a3#)
REM a1# = v#: a2# = r#: a4# = sr#
b3# = sr#
x# = (v# / 4) - INT(v# / 4)
x# = x# * 10
x# = INT(x#)
IF x# = 0 THEN z = 0
IF x# = 5 THEN z = 0
```

```
IF x# = 2 THEN z = 1

IF x# = 7 THEN z = 1

dist# = (v# - r#)

IF (dist# / 2) - INT(dist# / 2) = .5 THEN dist# = dist# - 1

dist# = dist# / 2

dist# = dist# + 1

IF z = 0 THEN dist# = dist# - 1

IF z = 1 THEN dist# = dist# - 2

d# = dist#

IF ((d# - sr#) = -1) THEN flag = 1 ELSE flag = 0

dist# = (v# - r#)

IF (dist# / 2) - INT(dist# / 2) = .5 THEN dist# = dist# - 1

dist# = dist# / 2

dist# = dist# + 1

b1# = v# - dist#

IF flag = 1 THEN b3# = b3# - 8

IF flag = 1 THEN GOTO skip2:

IF z = 1 THEN b3# = b3# - 8

skip2:

IF b3# < 0 THEN b3# = 0

b2# = r# - b3#

IF b2# < 1 THEN b2# = 1

test = (b1# = a1#) AND (b2# = a2#)

REM test = (b1# = a1#) AND (b2# = a2#) AND (b3# = a3#)
```

```
IF test = -1 THEN EXIT SUB

IF test = 0 THEN PRINT " "

REM IF test = 0 THEN PRINT b1#; a1#, b2#; a2#, b3#; a3#

IF test = 0 THEN PRINT b1#; a1#, b2#; a2#

IF test = 0 THEN STOP

END SUB
```

```
"C:\Users\Reactor1967\vcns\work\100403A.BAS"

REM Program works just fine until data is coded for some unknown reason

REM the data is messing things up when there is no reason for it to

REM mess things up. This program should be able to function with data.

REM PROGRAMERS NOTE CHANGE CODE POSITION FROM (D# - SR# = 0 TO D# - SR# = -1)

DECLARE SUB decodecheck (v#, r#, d#, sr#, a1#, a2#, a3#)

v# = 1

r# = 1

sr# = 0

RANDOMIZE TIMER

CLS

cracker = 0

DO

d# = v# - r#
```

```
REM IF d# = sr# THEN z = INT(RND * 2) + 0
burris = (d# = sr#)
IF z = 0 THEN v# = v# + (v# - r# + 1)
IF z = 1 THEN v# = v# + (v# - r# + 2)
m# = (v# / 4) - INT(v# / 4)
m# = m# * 10
m# = INT(m#)
PRINT z; m#; v#; r#; v# - r#; d#; sr#; d# - sr#; burris
IF cracker = 1 THEN CALL decodecheck(v#, r#, d#, sr#, a1#, a2#, a3#)
a1# = v#: a2# = r#: a3# = sr#
a$ = INKEY$
REM INPUT a$
IF a$ = "s" THEN STOP
IF a$ = "d" THEN EXIT DO
r# = r# + sr#
test = r# + (sr# + 7) <= v#
lb# = sr#
IF test = -1 THEN sr# = sr# + 8
IF test = 0 THEN z = 0
IF test = -1 THEN z = 1
cracker = 1
LOOP
DO
x# = (v# / 4) - INT(v# / 4)
```

```
x# = x# * 10
x# = INT(x#)
IF x# = 0 THEN z = 0
IF x# = 5 THEN z = 0
IF x# = 2 THEN z = 1
IF x# = 7 THEN z = 1
dist# = (v# - r#)
IF (dist# / 2) - INT(dist# / 2) = .5 THEN dist# = dist# - 1
dist# = dist# / 2
dist# = dist# + 1
IF z = 0 THEN dist# = dist# - 1
IF z = 1 THEN dist# = dist# - 2
d# = dist#
IF d# = sr# THEN flag = 1 ELSE flag = 0
burris = (d# = sr#)
m# = (v# / 4) - INT(v# / 4)
m# = m# * 10
m# = INT(m#)
PRINT z; m#; v#; r#; v# - r#; d#; sr#; d# - sr#; burris
a$ = INKEY$
REM INPUT a$
IF a$ = "s" THEN STOP
IF v# <= 1 THEN EXIT DO
dist# = (v# - r#)
```

```
IF (dist# / 2) - INT(dist# / 2) = .5 THEN dist# = dist# - 1

dist# = dist# / 2

dist# = dist# + 1

v# = v# - dist#

REM IF flag = 1 THEN b3# = b3# - 8

IF flag = 1 THEN GOTO skip:

IF z = 1 THEN sr# = sr# - 8

skip:

r# = r# - sr#

LOOP

PRINT "DECODE COMPLETE"


SUB decodecheck (v#, r#, d#, sr#, a1#, a2#, a3#)

REM a1# = v#: a2# = r#: a4# = sr#

b3# = sr#

x# = (v# / 4) - INT(v# / 4)

x# = x# * 10

x# = INT(x#)

IF x# = 0 THEN z = 0

IF x# = 5 THEN z = 0

IF x# = 2 THEN z = 1

IF x# = 7 THEN z = 1

dist# = (v# - r#)

IF (dist# / 2) - INT(dist# / 2) = .5 THEN dist# = dist# - 1
```

```
dist# = dist# / 2

dist# = dist# + 1

IF z = 0 THEN dist# = dist# - 1

IF z = 1 THEN dist# = dist# - 2

d# = dist#

IF d# = sr# THEN flag = 1 ELSE flag = 0

dist# = (v# - r#)

IF (dist# / 2) - INT(dist# / 2) = .5 THEN dist# = dist# - 1

dist# = dist# / 2

dist# = dist# + 1

b1# = v# - dist#

REM IF flag = 1 THEN b3# = b3# - 8

IF flag = 1 THEN GOTO skip2:

IF z = 1 THEN b3# = b3# - 8

skip2:

IF b3# < 0 THEN b3# = 0

b2# = r# - b3#

IF b2# < 1 THEN b2# = 1

test = (b1# = a1#) AND (b2# = a2#)

REM test = (b1# = a1#) AND (b2# = a2#) AND (b3# = a3#)

IF test = -1 THEN EXIT SUB

IF test = 0 THEN PRINT " "

REM IF test = 0 THEN PRINT b1#; a1#, b2#; a2#, b3#; a3#

IF test = 0 THEN PRINT b1#; a1#, b2#; a2#
```

```
IF test = 0 THEN STOP
END SUB
```

```
"C:\Users\Reactor1967\vcns\work\100203A.BAS"
v# = 1
r# = 1
sr# = 0
z = 1
CLS
RANDOMIZE TIMER
DO
IF z = 0 THEN v# = v# + (v# - r# + 1)
IF z = 1 THEN v# = v# + (v# - r# + 2)
m# = (v# / 4) - INT(v# / 4): m# = m# * 10: m# = INT(m#)
PRINT , z; m#; v#; r#; v# - r#; d#; sr#; d# - sr#
a$ = INKEY$
REM INPUT a$
IF a$ = "d" THEN EXIT DO
IF a$ = "s" THEN STOP
lb# = sr#
DO
sr# = sr# + 4
```

```
test = (sr# > INT((v# - r#) / 2))
IF test = -1 THEN EXIT DO
a$ = INKEY$: IF a$ = "s" THEN STOP
LOOP
sr# = sr# - 4
IF sr# = lb# THEN z = 0 ELSE z = 1
d# = (v# - r#) - sr#
r# = v# - d#
LOOP
REM DECODE ROUTINE
DO
x# = (v# / 4) - INT(v# / 4): x# = x# * 10: x# = INT(x#)
IF x# = 0 THEN z = 0
IF x# = 5 THEN z = 0
IF x# = 2 THEN z = 1
IF x# = 7 THEN x = 1
d# = (v# - r#)
IF (d# / 2) - INT(d# / 2) THEN d# = d# - 1
d# = d# / 2
d# = d# + 1
IF z = 0 THEN d# = d# - 1
IF z = 1 THEN d# = d# - 2
m# = (v# / 4) - INT(v# / 4): m# = m# * 10: m# = INT(m#)
PRINT , z; m#; v#; r#; v# - r#; d#; sr#; ABS(d# - sr#)
```

```
IF v# <= 1 THEN SYSTEM

a$ = INKEY$

REM INPUT a$

IF a$ = "s" THEN STOP

dist# = v# - r#

IF (dist# / 2) - INT(dist# / 2) = .5 THEN dist# = dist# - 1

dist# = dist# / 2

dist# = dist# + 1

v# = v# - dist#

r# = r# - sr#

IF r# < 1 THEN r# = 1

REM put sr# speed increase decrease here.

IF z = 1 THEN sr# = sr# - 4

LOOP
```

"C:\Users\Reactor1967\vcns\work\100103B.BAS"

```
REM This program does not work

v# = 300

r# = 1

CLS

RANDOMIZE TIMER

DO
```

```
z = INT(RND * 2) + 0
d# = (v# - r#) - 198
r# = v# - d#
IF z = 0 THEN v# = v# + ((v# - r#) * 2) + 1
IF z = 1 THEN v# = v# + ((v# - r#) * 2) + 2
IF d# >= 102 THEN flag = 1 ELSE flag = 0
IF flag = 1 THEN v# = v# - 16
PRINT , , z; v#; r#; v# - r#; d#; r# - sr#; flag
sr# = r#
a$ = INKEY$
INPUT a$
IF a$ = "s" THEN STOP
LOOP
```

"C:\Users\Reactor1967\vcns\work\100103A.BAS"

```
REM If I could just do this for base 3 then I could make this work
v# = 200
r# = 1
CLS
RANDOMIZE TIMER
DO
z = INT(RND * 2) + 0
```

```
d# = (v# - r#) - 100

r# = v# - d#

IF z = 0 THEN v# = v# + (v# - r#) + 1

IF z = 1 THEN v# = v# + (v# - r#) + 2

IF (v# - r#) >= 209 THEN flag = 1 ELSE flag = 0

IF (v# - r#) >= 209 THEN v# = v# - 10

PRINT , , z; v#; r#; v# - r#; d#; r# - sr#; flag

sr# = r#

a$ = INKEY$

REM INPUT a$

IF a$ = "s" THEN STOP

LOOP
```

"C:\Users\Reactor1967\vcns\work\092803B.BAS"

```
v# = 51

r# = 1

RANDOMIZE TIMER

x# = 0

REM does not work

CLS

DO

z = INT(RND * 2) + 0
```

```
IF z = 0 THEN v# = v# + ((v# - r#) * 2) + 1

IF z = 1 THEN v# = v# + ((v# - r#) * 2) + 2

IF z = 2 THEN v# = v# + ((v# - r#) * 2) + 3

PRINT , , z; v#; r#; v# - r#; d#; sd#

a$ = INKEY$

INPUT a$

IF a$ = "s" THEN STOP

d# = (v# - r#) - 100

sd# = d#

IF d# > 104 THEN d# = d# - 206

r# = v# - d#

LOOP
```

```
"C:\Users\Reactor1967\vcns\work\092803A.BAS"

REM does not work

DECLARE SUB code1 (v#, r#, z!, x#)

DECLARE SUB code2 (v#, r#, z!, x#)

v# = 201

r# = 1

CLS

x# = 8

DO
```

```
z = INT(RND * 2) + 0
REM z = 0
d# = (v# - r#) - 100
test1 = (d# >= 104)
test2 = (d# <= 103)
IF test1 = -1 THEN flag = 1
IF test2 = -1 THEN flag = 0
r# = v# - d#
sv# = v# - r#
IF flag = 1 THEN z = 2
IF flag = 1 THEN CALL code1(v#, r#, z, x#)
IF flag = 0 THEN CALL code2(v#, r#, z, x#)
PRINT z; v#; r#; v# - r#; sv#; r# - sr#; flag; d#
sr# = r#
a$ = INKEY$
INPUT a$
IF a$ = "s" THEN STOP
LOOP


SUB code1 (v#, r#, z, x#)
dist# = ((v# - r#) * 2) + z + 1 - 307
IF z = 0 THEN v# = v# + dist#
IF z = 1 THEN v# = v# + dist#
IF z = 2 THEN v# = v# + dist#
```

```
END SUB


SUB code2 (v#, r#, z, x#)
IF z = 0 THEN v# = v# + ((v# - r#) * 2) + 1
IF z = 1 THEN v# = v# + ((v# - r#) * 2) + 2
IF z = 2 THEN v# = v# + ((v# - r#) * 2) + 3
END SUB




"C:\Users\Reactor1967\vcns\work\092703C.BAS"
DECLARE SUB code1 (v#, r#, z!, x#)
DECLARE SUB code2 (v#, r#, z!, x#)
REM x# is the break and the accelerator to control the speed of r#
REM v1# = v2# - ((((v2# - r2#) +- x#) / 2) + 1) decoding
REM v2# = v1# -+ (((v# - r#) or (r# - v#)) * base-1) + N +-x#
v# = 201
r# = 1
CLS


lb = 0
DO
IF r# >= 301 THEN lb = lb + 1: IF lb = 5 THEN lb = 1
IF lb = 1 THEN z = INT(RND * 2) + 0
```

```
d# = (v# - r#) - 100
test1 = (d# >= 104)
test2 = (d# <= 103)
IF test1 = -1 THEN flag = 1
IF test2 = -1 THEN flag = 0
r# = v# - d#
IF flag = 1 THEN CALL code1(v#, r#, z, x#)
IF flag = 0 THEN CALL code2(v#, r#, z, x#)
PRINT z; v#; r#; v# - r#; r# - sr#; flag; d#; lb
sr# = r#
a$ = INKEY$
INPUT a$
IF a$ = "s" THEN STOP
LOOP

SUB code1 (v#, r#, z, x#)
REM dist# = (v# - r#) + z + 1 - 10
dist# = (v# - r#) + z + 1 - 10
IF z = 0 THEN v# = v# + dist#
IF z = 1 THEN v# = v# + dist#
END SUB

SUB code2 (v#, r#, z, x#)
IF z = 0 THEN v# = v# + (v# - r# + 1)
```

```
IF z = 1 THEN v# = v# + (v# - r# + 2)

END SUB
```

"C:\Users\Reactor1967\vcns\work\092703B.BAS"

```
DECLARE SUB code1 (v#, r#, z!, x#)

DECLARE SUB code2 (v#, r#, z!, x#)

REM v1# = v2# - ((((v# - r#)- x#) / 2) + 1) decoding if (v# - r#) is odd
minus 1

REM THIS PROGRAM WORKS YOU JUST NEED TO KNOW WHEN
TO DECODE FOR X AND

REM WHEN NOT TO DECODE FOR X.

v# = 201

r# = 1

CLS

x# = 8

DO

z = INT(RND * 2) + 0

REM z = 0

d# = (v# - r#) - 100

test1 = (d# >= 104)

test2 = (d# < 104)

IF test1 = -1 THEN flag = 1

IF test2 = -1 THEN flag = 0
```

```
r# = v# - d#

IF flag = 1 THEN CALL code1(v#, r#, z, x#)

IF flag = 0 THEN CALL code2(v#, r#, z, x#)

d2# = v# - r#

IF (d2# / 2) - INT(d2# / 2) = .5 THEN d2# = d2# - 1

d2# = d2# / 2: d2# = d2# + 1: d2# = d2# - z - 1

PRINT z; v#; r#; v# - r#; r# - sr#; flag; d#; d2#

sr# = r#

a$ = INKEY$

REM INPUT a$

IF a$ = "s" THEN STOP

LOOP


SUB code1 (v#, r#, z, x#)

dist# = (v# - r#) + z + 1 - 10

IF z = 0 THEN v# = v# + dist#

IF z = 1 THEN v# = v# + dist#

END SUB


SUB code2 (v#, r#, z, x#)

IF z = 0 THEN v# = v# + (v# - r# + 1)

IF z = 1 THEN v# = v# + (v# - r# + 2)

END SUB
```

```
"C:\Users\Reactor1967\vcns\work\092703A.BAS"
v# = 1
r# = 1
x# = 0
CLS
RANDOMIZE TIMER
DO
z = INT(RND * 2) + 0
DO
dist# = x#: x# = x# + 1
IF (dist# / 2) - INT(dist# / 2) = .5 THEN dist# = dist# - 1
dist# = dist# / 2: dist# = dist# + 1
IF z = 0 THEN dist# = dist# - 1: IF z = 1 THEN dist# = dist# - 2
r2# = v# - dist#
test1 = INT((((r# - r2#) / 4) - INT((r# - r2#) / 4)) * 10) = 0
test3 = (z = 0) AND ((x# / 2) - INT(x# / 2) = .5)
test4 = (z = 1) AND ((x# / 2) - INT(x# / 2) = 0)
test5 = (test3 = -1) OR (test4 = -1)
test6 = (test1 = -1) AND (test5 = -1)
IF test6 = -1 THEN EXIT DO
a$ = INKEY$: IF a$ = "s" THEN STOP
LOOP
```

```
r# = r2#
IF z = 0 THEN v# = v# + ((v# - r#) * 1) + 1
IF z = 1 THEN v# = v# + ((v# - r#) * 1) + 2
m# = INT((((v#) / 4) - INT((v#) / 4)) * 10)
PRINT z; m#; v#; r#; v# - r#; x#
svr# = (v# - r#)
a$ = INKEY$
IF a$ = "s" THEN STOP
IF a$ = "d" THEN EXIT DO
LOOP
```

```
"C:\Users\Reactor1967\vcns\work\092603E.BAS"
REM when you use x# as a throttle for your v# - r# you can find N
REM by dividing (v# - r#) by the base.
v# = 3
r# = 1
x# = 0
CLS
RANDOMIZE TIMER
DO
z = INT(RND * 3) + 0
REM IF z = 0 THEN x# = x# + 1
```

```
REM IF z = 1 THEN x# = x# + 2

REM IF z = 2 THEN x# = x# + 3

x# = x# + 1

r# = v# - x#

IF z = 0 THEN v# = v# + ((v# - r#) * 2) + 1

IF z = 1 THEN v# = v# + ((v# - r#) * 2) + 2

IF z = 2 THEN v# = v# + ((v# - r#) * 2) + 3

m# = INT(((v# / 3) - INT(v# / 3)) * 10)

PRINT z; m#; v#; r#; v# - r#; x#; x# - sx#; INT((((v# - r#) / 3) - INT((v# - r#) / 3)) * 10)

sx# = x#

a$ = INKEY$

IF a$ = "s" THEN STOP

IF a$ = "d" THEN EXIT DO

LOOP
```

```
"C:\Users\Reactor1967\vcns\work\092603D.BAS"

v# = 3

r# = 1

x# = 0

CLS

RANDOMIZE TIMER
```

```
DO

z = INT(RND * 2) + 0

IF z = 0 THEN x# = x# + 1

IF z = 1 THEN x# = x# + 2

r# = v# - x#

IF z = 0 THEN v# = v# + ((v# - r#) * 1) + 1

IF z = 1 THEN v# = v# + ((v# - r#) * 1) + 2

PRINT z; v#; r#; v# - r#; x#; x# - sx#; INT((((v# - r#) / 4) - INT((v# -
r#) / 4)) * 10)

sx# = x#

a$ = INKEY$

IF a$ = "s" THEN STOP

IF a$ = "d" THEN EXIT DO

LOOP
```

```
"C:\Users\Reactor1967\vcns\work\092603C.BAS"

v# = 1

r# = 1

x# = 0

CLS

RANDOMIZE TIMER

DO
```

```
z = INT(RND * 2) + 0

z1 = (v# / 2) - INT(v# / 2)

IF z1 = .5 THEN z1 = 1

IF z = 0 THEN v# = v# + ((v# - r#) * 1) + 1

IF z = 1 THEN v# = v# + ((v# - r#) * 1) + 2

PRINT z; v#; r#; v# - r#; x#; x# - sx#

sx# = x#

a$ = INKEY$

IF a$ = "s" THEN STOP

IF a$ = "d" THEN EXIT DO

z2 = (v# / 2) - INT(v# / 2)

IF z2 = .5 THEN z2 = 1

test1 = (z1 = 0) AND (z2 = 0)

test2 = (z1 = 0) AND (z2 = 1)

test3 = (z1 = 1) AND (z2 = 0)

test4 = (z1 = 1) AND (z2 = 1)

IF test1 = -1 THEN x# = x# + 2

IF test2 = -1 THEN x# = x# + 3

IF test3 = -1 THEN x# = x# + 3

IF test4 = -1 THEN x# = x# + 4

r# = v# - x#

LOOP
```

```
"C:\Users\Reactor1967\vcns\work\092603B.BAS"
v# = 1
r# = 1
x# = 0
CLS
RANDOMIZE TIMER
REM 0 to 1 x# = x# + 3
REM 1 to 0 x# = x# + 1
z1 = 1
DO
z = INT(RND * 2) + 0
IF z = 0 THEN v# = v# + ((v# - r#) * 1) + 1
IF z = 1 THEN v# = v# + ((v# - r#) * 1) + 2
PRINT z; v#; r#; v# - r#; x#; x# - sx#
sx# = x#
a$ = INKEY$
IF a$ = "s" THEN STOP
IF a$ = "d" THEN EXIT DO
test1 = (z1 = 1) AND (z = 0)
test2 = (z1 = 1) AND (z = 1)
test3 = (z1 = 0) AND (z = 0)
test4 = (z1 = 0) AND (z = 1)
IF test1 = -1 THEN x# = x# + 1
```

```
IF test2 = -1 THEN x# = x# + 4

IF test3 = -1 THEN x# = x# + 4

IF test4 = -1 THEN x# = x# + 3

r# = v# - x#

z1 = (v# / 2) - INT(v# / 2)

IF z1 = .5 THEN z1 = 1

LOOP
```

```
"C:\Users\Reactor1967\vcns\work\092603A.BAS"

v# = 1

r# = 1

x# = 0

CLS

RANDOMIZE TIMER

REM 0 to 1 x# = x# + 3

REM 1 to 0 x# = x# + 1

z = 1

DO

l = z

z = INT(RND * 2) + 0

IF z = 0 THEN v# = v# + ((v# - r#) * 1) + 1

IF z = 1 THEN v# = v# + ((v# - r#) * 1) + 2
```

```
PRINT z; v#; r#; v# - r#; x#; x# - sx#
sx# = x#
a$ = INKEY$
IF a$ = "s" THEN STOP
IF a$ = "d" THEN EXIT DO
test1 = (l = 0) AND (z = 0)
test2 = (l = 0) AND (z = 1)
test3 = (l = 1) AND (z = 0)
test4 = (l = 1) AND (z = 1)
IF test1 = -1 THEN x# = x# + 4
IF test2 = -1 THEN x# = x# + 3
IF test3 = -1 THEN x# = x# + 1
IF test4 = -1 THEN x# = x# + 4
r# = v# - x#
LOOP
l = (v# / 2) - INT(v# / 2)
IF l = .5 THEN z1 = 1
dist# = v# - r#
IF (dist# / 2) - INT(dist# / 2) = .5 THEN dist# = dist# - 1
dist# = dist# / 2
dist# = dist# + 1
v# = v# - dist#
z = (v# / 2) - INT(v# / 2)
IF z = .5 THEN z2 = 1
```

```
test1 = (l = 0) AND (z = 0)
test2 = (l = 0) AND (z = 1)
test3 = (l = 1) AND (z = 0)
test4 = (l = 1) AND (z = 1)
IF test1 = -1 THEN x# = x# - 4
IF test2 = -1 THEN x# = x# - 3
IF test3 = -1 THEN x# = x# - 1
IF test4 = -1 THEN x# = x# - 4
d# = x#
d# = d# * 2
IF z = 0 THEN d# = d# + 1
IF z = 1 THEN d# = d# + 2
r# = v# - d#
PRINT , , v#; r#; v# - r#
```

"C:\Users\Reactor1967\vcns\work\092403A.BAS"

```
DIM can1(1000)
DIM can2(1000)
DIM can3(1000)
redo:
v# = 1
r# = 1
```

```
x# = 0
RANDOMIZE TIMER
CLS
DO
z = INT(RND * 2) + 0
dg# = (v# - r#)
IF z = 0 THEN v# = v# + (v# - r# + 1)
IF z = 1 THEN v# = v# + (v# - r# + 2)
m# = (v# / 4) - INT(v# / 4): m# = m# * 10: m# = INT(m#)
PRINT z; m#; v#; r#; v# - r#; dg#; m2#; (x# - 1) - sx#; v# - sv#; r# - sr#
IF dg# >= 1321 THEN STOP
sr# = r#
a$ = INKEY$
IF a$ = "s" THEN STOP
sv# = v#
sx# = x# - 1
m2# = ((v# - r#) / 4) - INT((v# - r#) / 4): m2# = m2# * 10: m2# =
INT(m2#)
dist# = (v# - r#)
dist# = INT(dist# / 2)
DO
s# = (v# - r#) - x#
IF (s# / 4) - INT(s# / 4) = 0 THEN EXIT DO
x# = x# + 1
```

```
a$ = INKEY$
IF a$ = "s" THEN STOP
LOOP
r# = v# - x#
x# = x# + 1
LOOP
```

"C:\Users\Reactor1967\vcns\work\092203E.BAS"

```
v# = 1
r# = 1
CLS
RANDOMIZE TIMER
DO
z = INT(RND * 2) + 0
x# = (v# - r#)
IF z = 0 THEN v# = v# + ((v# - r#) * 1) + 1
IF z = 1 THEN v# = v# + ((v# - r#) * 1) + 2
m# = (v# / 4) - INT(v# / 4): m# = m# * 10: m# = INT(m#)
j# = v# - r#: m2# = (j# / 4) - INT(j# / 4): m2# = m2# * 10: m2# = INT(m2#)
m3# = (x# / 4) - INT(x# / 4): m3# = m3# * 10: m3# = INT(m3#)
PRINT , z; m#; v#; r#; v# - r#; x#; m2#; m3#
```

```
a$ = INKEY$
IF a$ = "s" THEN STOP
dist# = v# - r#
dist# = INT(dist# / 4)
dist# = dist# * 10
dist# = INT(dist#)
ab# = 1
DO
d# = INT(ab# / 4) - INT(ab# / 4)
d# = d# * 10
d# = INT(d#)
IF d# = dist# THEN EXIT DO
ab# = ab# + 1
a$ = INKEY$: IF a$ = "s" THEN STOP
LOOP
r# = v# - ab#
LOOP
```

```
"C:\Users\Reactor1967\vcns\work\092203D.BAS"
v# = 1
r# = 1
CLS
```

```
RANDOMIZE TIMER

DO

z = INT(RND * 2) + 0

x# = (v# - r#)

IF z = 0 THEN v# = v# + ((v# - r#) * 1) + 1

IF z = 1 THEN v# = v# + ((v# - r#) * 1) + 2

m# = (v# / 4) - INT(v# / 4): m# = m# * 10: m# = INT(m#)

j# = v# - r#: m2# = (j# / 4) - INT(j# / 4): m2# = m2# * 10: m2# =
INT(m2#)

m3# = (x# / 4) - INT(x# / 4): m3# = m3# * 10: m3# = INT(m3#)

PRINT , z; m#; v#; r#; v# - r#; x#; m2#; m3#

a$ = INKEY$

IF a$ = "s" THEN STOP

dist# = INT((v# - r#) / 2)

DO

s# = (v# - r#) - dist#

test = (s# / 4) - INT(s# / 4) = 0

IF test = -1 THEN EXIT DO

dist# = dist# + 1

a$ = INKEY$: IF a$ = "s" THEN STOP

LOOP

r# = v# - dist#

LOOP
```

```
"C:\Users\Reactor1967\vcns\work\092203C.BAS"

d# = 0

CLS

DO

x# = INT(d# / 2)

DO

s# = d# - x#

IF (s# / 3) - INT(s# / 3) = 0 THEN EXIT DO

x# = x# + 1

a$ = INKEY$: IF a$ = "s" THEN STOP

LOOP

PRINT , , d#; x#; s#

a$ = INKEY$: INPUT a$: IF a$ = "s" THEN STOP

d# = d# + 1

LOOP
```

```
"C:\Users\Reactor1967\vcns\work\092203B.BAS"

DECLARE SUB dist (v#, r#)

v# = 100
```

```
r# = 1

CLS

RANDOMIZE TIMER

DO

z = INT(RND * 2) + 0

CALL dist(v#, r#)

IF z = 0 THEN v# = v# + ((v# - r#) * 2) + 1

IF z = 1 THEN v# = v# + ((v# - r#) * 2) + 2

IF z = 2 THEN v# = v# + ((v# - r#) * 2) + 3

PRINT , , z; v#; r#; v# - r#

a$ = INKEY$

IF a$ = "s" THEN STOP

LOOP


SUB dist (v#, r#)

d# = 6

DO

x# = INT(d# / 2)

DO

s# = d# - x#

IF (s# / 3) - INT(s# / 3) = 0 THEN EXIT DO

x# = x# + 1

a$ = INKEY$: IF a$ = "s" THEN STOP

LOOP
```

```
REM PRINT , , d#; x#; s#

a$ = INKEY$: REM  INPUT a$: IF a$ = "s" THEN STOP

IF d# = (v# - r#) THEN EXIT DO

d# = d# + 1

LOOP

r# = v# - x#

END SUB
```

"C:\Users\Reactor1967\vcns\work\092203A.BAS"

```
REM This program can take v# & r# and switch back and from base 2 to
base

REM 3 and back again.

REM since sysmetery is directly to the speed of r# all one must do is have

REM the speed of r# divisable by the base with no remainder or a
constant

REM speed of r# to acheive sysmetery.

REM USE THIS IN BOOK - WHAT BASE YOU CAN USE AND KEEP
SYSMETERY DEPENDS

REM ON THE SPEED THAT YOUR REFERENCE POINT IS
TRAVELING.

REM its possible to determine what d#'s(d# = v# - r# before coding that
comes

REM from a specific v# - r# after coding) come from what specific v# -
r#
```

```
REM I thing a alogermithm(misspelled) might be in order here. Knowing
what

v# = 1

r# = 1

CLS

RANDOMIZE TIMER

REM GOTO skip:

redo:

DO

z = INT(RND * 2) + 0

x# = v# - r#

IF z = 0 THEN v# = v# + (v# - r# + 1)

IF z = 1 THEN v# = v# + (v# - r# + 2)

m# = (v# / 4) - INT(v# / 4): m# = m# * 10: m# = INT(m#)

PRINT z; m#; v#; r#; v# - r#; x#; r# - sr#; "2"; (v# / 3) - INT(v# / 3)

sr# = r#

a$ = INKEY$

IF a$ = "s" THEN STOP

IF a$ = "d" THEN EXIT DO

dist# = INT((v# - r#) / 2)

DO

s# = (v# - r#) - dist#

IF (s# / 4) - INT(s# / 4) = 0 THEN EXIT DO

dist# = dist# + 1
```

```
a$ = INKEY$: IF a$ = "s" THEN STOP
LOOP
r# = v# - dist#
LOOP
REM ----------------------------
skip:
DO
z = INT(RND * 3) + 0
x# = v# - r#
IF z = 0 THEN v# = v# + ((v# - r#) * 2) + 1
IF z = 1 THEN v# = v# + ((v# - r#) * 2) + 2
IF z = 2 THEN v# = v# + ((v# - r#) * 2) + 3
m# = (v# / 3) - INT(v# / 3): m# = m# * 10: m# = INT(m#)
PRINT z; m#; v#; r#; v# - r#; x#; r# - sr#; "3"; (v# / 2) - INT(v# / 2)
sr# = r#
a$ = INKEY$
REM INPUT a$
IF a$ = "s" THEN STOP
IF a$ = "d" THEN EXIT DO
dist# = INT((v# - r#) / 3)
DO
s# = (v# - r#) - dist#
IF (s# / 3) - INT(s# / 3) = 0 THEN EXIT DO
dist# = dist# + 1
```

```
a$ = INKEY$: IF a$ = "s" THEN STOP

LOOP

r# = v# - dist#

LOOP

GOTO redo:
```

```
"C:\Users\Reactor1967\vcns\work\092003A.BAS"

REM New equations and formula's comming here. So ever v# - r# had a d#

REM if working base two then there are two d's bot the same

REM for each N. If working base three then there are three d's

REM each the same for each value of N. so d# is the same for each value

REM on N for the base. Draw a chart showing which d# = (v# - r#) before coding

REM that came immediately after (v# - r#) after coding. Keep the speed of r#

REM in check.

REM -----------

REM conclusions of experiments

REM Having the same d# for two (v# - r#) is very difficult. It seems to keep

REM sysmetery that the d# has to go for two N's of the same value. Now this
```

```
REM is possible but only if various ranges of (v# - r#)'s can be applied. The

REM reason is there is no way to know which v# - r# the d# came from. This has

REM being the paradox I have run into time and time again working with this

REM numerical system that each answer always seems to create two or more other

REM problems that have to be solved thus always ending in a loop some where.

REM But this here does seem promising and interesting to work with.

DIM can1(1000)

DIM can2(1000)

DIM can3(1000)

redo:

a# = v# - r#

s# = 0

v# = 1

r# = 1

CLS

DO

z = INT(RND * 2) + 0

IF z = 0 THEN v# = v# + (v# - r# + 1)

IF z = 1 THEN v# = v# + (v# - r# + 2)

m# = (v# / 4) - INT(v# / 4)

m# = m# * 10
```

```
m# = INT(m#)

PRINT , z; m#; v#; r#; v# - r#; x#; r# - sr#

a$ = INKEY$

REM INPUT a$

IF a$ = "s" THEN STOP

IF a$ = "d" THEN GOTO ou:

IF a# > 1000 THEN GOTO redo:

can1(a#) = z2

can2(a#) = (a#)

can3(a#) = x#

sr# = r#

dist# = INT((v# - r#) / 2)

z2 = z

a# = v# - r#

r# = v# - dist#

x# = v# - r#

LOOP

ou:

FOR count = 1 TO 1000

PRINT , , can1(count); can2(count); can3(count)

a$ = INKEY$

REM INPUT a$

IF a$ = "s" THEN STOP

NEXT count
```

```
"C:\Users\Reactor1967\vcns\work\091403A.BAS"

REM Here the distance coded to is predetermined before each code.

REM You can code to any distance you want to code to the trick is

REM doing it in a way you can decode by knowing what your next v - r

REM is or the speed of r. I still have not got a clue how to do that.

REM ---------------------------------------------------------------

REM This may be the first step of knowing how to figure your distance
and

REM v# - r# when decoding. Study the pattern here. It might be useful.

REM Opening up something new here.

DIM can1(1000)

DIM can2(1000)

DIM can3(1000)

FOR count = 0 TO 1000

can1(count) = count

NEXT count

store# = 0

redo2:

d# = 0

v# = 1000

CLS
```

```
RANDOMIZE TIMER
DO
z = INT(RND * 2) + 0
redo:
test = (d# / 2) - INT(d# / 2)
IF test = 0 THEN z2 = 1
IF test = .5 THEN z2 = 0
IF z <> z2 THEN flag = 1 ELSE flag = 0
IF flag = 1 THEN d# = d# + 1
a$ = INKEY$: IF a$ = "s" THEN STOP
IF flag = 1 THEN GOTO redo:
REM PRINT d#; z
dist# = d#
IF (dist# / 2) - INT(dist# / 2) = .5 THEN dist# = dist# - 1
dist# = dist# / 2
dist# = dist# + 1
IF z = 0 THEN dist# = dist# - 1
IF z = 1 THEN dist# = dist# - 2
r# = v# - dist#
REM IF (r# / 2) - INT(r# / 2) = 0 THEN d# = d# + 1
a$ = INKEY$: IF a$ = "s" THEN STOP
REM IF (r# / 2) - INT(r# / 2) = 0 THEN GOTO redo:
test = (r# / 4) - INT(r# / 4)
IF test <> .25 THEN d# = d# + 1
```

```
IF test <> .25 THEN GOTO redo:

IF z = 0 THEN v# = v# + (v# - r# + 1)

IF z = 1 THEN v# = v# + (v# - r# + 2)

m# = (v# / 4) - INT(v# / 4): m# = m# * 10: m# = INT(m#)

yu# = (v# - r#)

m2# = (yu# / 4) - INT(yu# / 4): m2# = m2# * 10: m2# = INT(m2#)

PRINT z; m#; m2#; v#; r#; v# - r#; d#; (v# - r#) - svr#; v# - sv#; r# - sr#;
(r# / 4) - INT(r# / 4); ((r# - sr#) / 4) - INT((r# - sr#) / 4)

IF z = 0 THEN can2(store2#) = (v# - r#)

IF z = 1 THEN can3(store2#) = (v# - r#)

IF (v# - r#) >= 1000 THEN GOTO redo2:

store2# = (v# - r#)


IF v# - r# > 1000 THEN GOTO redo2:

sr# = r#: svr# = v# - r#: sv# = v#

IF (v# - r#) <> d# THEN STOP

a$ = INKEY$

REM INPUT a$

IF a$ = "s" THEN STOP

IF a$ = "d" THEN EXIT DO

d# = d# + 1

LOOP

FOR count = 1 TO 1000

PRINT , , can1(count); " "; can2(count); can3(count)
```

```
INPUT a$
IF a$ = "s" THEN STOP
NEXT count
```

"C:\Users\Reactor1967\vcns\work\091203A.BAS"

```
REM Here the distance coded to is predetermined before each code.
REM You can code to any distance you want to code to the trick is
REM doing it in a way you can decode by knowing what your next v - r
REM is or the speed of r. I still have not got a clue how to do that.
d# = 100
v# = 1000
CLS
RANDOMIZE TIMER
DO
z = INT(RND * 2) + 0
redo:
test = (d# / 2) - INT(d# / 2)
IF test = 0 THEN z2 = 1
IF test = .5 THEN z2 = 0
IF z <> z2 THEN flag = 1 ELSE flag = 0
IF flag = 1 THEN d# = d# + 1
a$ = INKEY$: IF a$ = "s" THEN STOP
```

```
IF flag = 1 THEN GOTO redo:

REM PRINT d#; z

dist# = d#

IF (dist# / 2) - INT(dist# / 2) = .5 THEN dist# = dist# - 1

dist# = dist# / 2

dist# = dist# + 1

IF z = 0 THEN dist# = dist# - 1

IF z = 1 THEN dist# = dist# - 2

r# = v# - dist#

REM IF (r# / 2) - INT(r# / 2) = 0 THEN d# = d# + 1

a$ = INKEY$: IF a$ = "s" THEN STOP

REM IF (r# / 2) - INT(r# / 2) = 0 THEN GOTO redo:

test = (r# / 4) - INT(r# / 4)

IF test <> .25 THEN d# = d# + 1

IF test <> .25 THEN GOTO redo:

IF z = 0 THEN v# = v# + (v# - r# + 1)

IF z = 1 THEN v# = v# + (v# - r# + 2)

m# = (v# / 4) - INT(v# / 4): m# = m# * 10: m# = INT(m#)

yu# = (v# - r#)

m2# = (yu# / 4) - INT(yu# / 4): m2# = m2# * 10: m2# = INT(m2#)

PRINT z; m#; m2#; v#; r#; v# - r#; d#; (v# - r#) - svr#; v# - sv#; r# - sr#;
(r# / 4) - INT(r# / 4); ((r# - sr#) / 4) - INT((r# - sr#) / 4)

sr# = r#: svr# = v# - r#: sv# = v#

IF (v# - r#) <> d# THEN STOP
```

```
a$ = INKEY$

REM INPUT a$

IF a$ = "s" THEN STOP

d# = d# + 1

LOOP
```

"C:\Users\Reactor1967\vcns\work\090703B.BAS"

```
REM Learning to decode by testing for r# a new procedure.

REM make this decode

redo:

v# = 1

r# = 1

RANDOMIZE TIMER

CLS

DO

cd# = v# - r#

z = INT(RND * 2) + 0

IF z = 0 THEN v# = v# + (v# - r# + 1)

IF z = 1 THEN v# = v# + (v# - r# + 2)

PRINT z; v#; r#; v# - r#; cd#; r# - sr#

sr# = r#

a$ = INKEY$
```

```
IF a$ = "d" THEN GOTO decode:

IF a$ = "s" THEN STOP

dist# = INT(.5 * (v# - r#))

r# = v# - dist#

LOOP

decode:

PRINT "-------------------------------"

DO

sv2# = v#

ty# = v# - r#

IF (ty# / 2) - INT(ty# / 2) = .5 THEN z = 0

IF (ty# / 2) - INT(ty# / 2) = 0 THEN z = 1

PRINT z; v#; r#; v# - r#

a$ = INKEY$

INPUT a$

IF a$ = "s" THEN STOP

dist# = (v# - r#)

IF (dist# / 2) - INT(dist# / 2) = .5 THEN dist# = dist# - 1

dist# = dist# / 2

dist# = dist# + 1

v# = v# - dist#

IF z = 0 THEN dist# = dist# - 1

IF z = 1 THEN dist# = dist# - 2

a# = dist#
```

```
DO
d2# = dist# + a#
r2# = v# - d2#
test = (r# = (r2# + a#)) AND ((v# + (v# - r2#) + z + 1) = sv#)
IF test = -1 THEN EXIT DO
a# = a# + 1
a$ = INKEY$
IF a$ = "s" THEN STOP
LOOP
r# = v# - (a# + dist#)
LOOP
```

```
"C:\Users\Reactor1967\vcns\work\090703A.BAS"
REM Learning to decode by testing for r# a new procedure.
redo:
v# = 1
r# = 1
RANDOMIZE TIMER
CLS
ti# = INT(RND * 10000) + 100
ti2# = 0
DO
```

```
ti2# = ti2# + 1

cd# = v# - r#: sv# = v#

z = INT(RND * 2) + 0

IF z = 0 THEN v# = v# + (v# - r# + 1)

IF z = 1 THEN v# = v# + (v# - r# + 2)

PRINT z; v#; r#; v# - r#; cd#; r# - sr#

sz = z

sr# = r#

a$ = INKEY$

IF a$ = "s" THEN STOP

IF a$ = "d" THEN GOTO decode:

IF ti2# >= ti# THEN GOTO decode:

r# = r# + INT((v# - r#) / 2)

IF (r# / 2) - INT(r# / 2) = 0 THEN r# = r# - 1

LOOP

decode:

sv2# = v#

z = (v# / 2) - INT(v# / 2)

dist# = (v# - r#)

IF (dist# / 2) - INT(dist# / 2) = .5 THEN dist# = dist# - 1

dist# = dist# / 2

dist# = dist# + 1

v# = v# - dist#

IF z = .5 THEN z = 1
```

```
IF z <> sz THEN STOP

IF v# <> sv# THEN STOP

IF z = 0 THEN dist# = dist# - 1

IF z = 1 THEN dist# = dist# - 2

IF dist# <> cd# THEN PRINT cd#; dist#

IF dist# <> cd# THEN STOP

a# = dist#

FOR count = 1 TO 25

r2# = v# - (dist# + a#)

r2# = r2# + a#

IF r2# = r# THEN GOTO test:

a# = a# - 1

NEXT count

STOP

test:

r2# = v# - (dist# + a#)

v2# = v# + (v# - r# + z + 1)

IF v2# = sv2# THEN GOTO redo:

SYSTEM
```

```
"C:\Users\Reactor1967\vcns\work\090603A.BAS"

REM I believe it is possible to create a chart where if the intersection
```

REM of two different v#'s gives a specific speed of r#. This can be done I

REM believe by two different means. One by knowing specific math rules not

REM to break a chart can be created or by starting a your very first v# then

REM create a chart by coding each and every v# to every number in its base.

REM when you get enough numbers reverse your plus and minus signs put r#

REM above v# and create a chart coding v# down in value. reversing charts

REM can be a trick process you may have to keep two different r#'s one for

REM coding up and one for coding down. Keep both r#'s handy when coding and

REM decoding.

REM --------------------

REM it might be possible to take v1 after decode and test for r1. Do the same

REM calculations as if going to code to v2 and calculate speed of r#. When

REM you get your current r# you may have found your correct speed of r# and

REM correct v# - r# for v1. You would make sure to do the same calculations

REM decoding that was done coding to make this work.

REM ----

```
REM to find r1 from r2 decode v1 from v2. Find your d. Then starting at
a
REM good specific r# test a different r# using d to find the speed of r# till
REM you get a r# that matches r2# anda v# that matches v2#
v# = 1
r# = 1
low = 9999999
high = 0
RANDOMIZE TIMER
CLS
DO
z = INT(RND * 2) + 0
sv# = v#
IF z = 0 THEN v# = v# + (v# - r#) + 1
IF z = 1 THEN v# = v# + (v# - r#) + 2
PRINT z; v#; r#; v# - r#; d#
REM PRINT z; v#; r#; v# - r#; v# - sv#; d#; low; high
IF (v# - sv#) - d# > high THEN high = (v# - sv#) - d#
IF (v# - sv#) - d# < low THEN low = (v# - sv#) - d#
a$ = INKEY$
IF a$ = "s" THEN STOP
r# = r# + d#
d# = v# - r#
IF (d# / 2) - INT(d# / 2) = .5 THEN d# = d# - 1
```

```
LOOP

"C:\Users\Reactor1967\vcns\work\090503B.BAS"

REM New eqation the speed of r# is roughly equal to (v# - r#) * (base -
1)

REM If its less than that (v# - r#) increases if its greater (v# - r#)

REM decreases.

REM if you keep your speed of r# divided by the base constant then you
will

REM always be able to divied your v# by the base to find your N.

CLS

d# = 0

s# = 0

v# = 1

r# = 1

RANDOMIZE TIMER

DO

z = INT(RND * 3) + 0

s1# = v#

IF z = 0 THEN v# = v# + ((v# - r#) * 2) + 1

IF z = 1 THEN v# = v# + ((v# - r#) * 2) + 2

IF z = 2 THEN v# = v# + ((v# - r#) * 2) + 3
```

```
s2# = v#

m# = (v# / 3) - INT(v# / 3)

m# = m# * 10

m# = INT(m#)

PRINT z; m#; v#; r#; v# - r#; s2# - s1#; (s2# - s1#) - sv#; r# - sr#; (r# - sr#) - sr2#

sv# = s2# - s1#

sr2# = (r# - sr#)

sr# = r#

a$ = INKEY$

REM INPUT a$

IF a$ = "s" THEN STOP

r# = r# + d#

DO

m# = (r# / 3) - INT(r# / 3)

m# = m# * 10

m# = INT(m#)

IF m# = 3 THEN EXIT DO

r# = r# - 1

a$ = INKEY$

IF a$ = "s" THEN STOP

LOOP

d# = (v# - r#) * 2

LOOP
```

```
"C:\Users\Reactor1967\vcns\work\090503A.BAS"
REM Looking at the speed of r# I believe it is possible
REM to setup some kind of system for looking a N and being able
REM to tell what to do with the speed of r#.
CLS
s# = 0
v# = 1
r# = 1
RANDOMIZE TIMER
DO
z = INT(RND * 2) + 0
IF z = 0 THEN v# = v# + (v# - r#) + 1
IF z = 1 THEN v# = v# + (v# - r#) + 2
PRINT , z; v#; r#; v# - r#; d#; s#
a$ = INKEY$
INPUT a$
IF a$ = "s" THEN STOP
DO
IF (s# + 2) >= INT((v# - r#) / 2) THEN EXIT DO
s# = s# + 2
d# = (v# - r#) - s#
```

```
a$ = INKEY$
IF a$ = "s" THEN STOP
LOOP
r# = r# + s#
LOOP
```

"C:\Users\Reactor1967\vcns\work\083103B.BAS"

```
REM How to get around my equation wall. Say you want to do base
REM 3 but with base 1 equations meaning that my equations are usually
REM like this v# = v# + ((v# - r#) * base - 1) + N
REM but in base 3 and above its hard to get r# to keep up with v#
REM unless your smarter than I am which is not an impossiblility.
REM so, code like your coding in base 2 but know what to add from
REM from each N in base to to get base 3 values. Also your using
REM base 3 not base to so n = 1 to 3 instead of n = 1 to 2
REM now I will see if I can do this with out my helper sub.
v# = 1
r# = 1
CLS
RANDOMIZE TIMER
REM 0 = 6
REM 1 = 0
```

```
REM 2 = 3

DO

z = INT(RND * 2) + 0

test = (r# + 96) <= v# + ((v# - r#) * 2)

IF test = -1 THEN z = 2

IF z = 0 THEN v# = v# + ((v# - r#) * 1) + 1

IF z = 1 THEN v# = v# + ((v# - r#) * 1) + 2

IF z = 2 THEN v# = v# + ((v# - r#) * 1) + 3

way# = v#

DO

m# = (v# / 3) - INT(v# / 3)

m# = m# * 10

m# = INT(m#)

test1 = (z = 0) AND (m# = 6)

test2 = (z = 1) AND (m# = 0)

test3 = (z = 2) AND (m# = 3)

IF test1 = -1 THEN EXIT DO

IF test2 = -1 THEN EXIT DO

IF test3 = -1 THEN EXIT DO

v# = v# + 1

a$ = INKEY$

IF a$ = "s" THEN STOP

LOOP

way# = v# - way#
```

```
PRINT , , z; m#; v#; r#; v# - r#; way#
a$ = INKEY$
INPUT a$
IF a$ = "s" THEN STOP
REM IF z = 2 THEN r# = r# + 93
LOOP
```

"C:\Users\Reactor1967\vcns\work\083103A.BAS"

```
REM -------------------------TEMPLATE------------------------------
REM DO NOT ALTER THIS PROGRAM IT MIGHT BE USEFUL AS A TEMPLATE FOR
REM OTHER PROGRAMS!!!!!!!!!!!!!!!!!!!!
REM How to get around my equation wall. Say you want to do base
REM 3 but with base 1 equations meaning that my equations are usually
REM like this v# = v# + ((v# - r#) * base - 1) + N
REM but in base 3 and above its hard to get r# to keep up with v#
REM unless your smarter than I am which is not an impossiblility.
REM so, code like your coding in base 2 but know what to add from
REM from each N in base to to get base 3 values. Also your using
REM base 3 not base to so n = 1 to 3 instead of n = 1 to 2
REM now I will see if I can do this with out my helper sub.
REM when using a N to tell you when to increment use your lowest N
```

```
REM not the highest N

REM 0 to a 0 add 1

REM 0 to a 1 add 1

REM 0 to a 2 add 1

REM 1 to a 0 add 2

REM 1 to a 1 add 2

REM 1 to a 2 add 2

REM 2 to a 0 add 0

REM 2 to a 1 add 0

REM 2 to a 2 add 0

v# = 1

r# = 1

CLS

RANDOMIZE TIMER

REM 0 = 6

REM 1 = 0

REM 2 = 3

DO

z = INT(RND * 2) + 0

z = z + 1

IF z = 0 THEN z = 1

IF flag = 1 THEN z = 0

IF z = 0 THEN v# = v# + ((v# - r#) * 1) + 1

IF z = 1 THEN v# = v# + ((v# - r#) * 1) + 2
```

```
IF z = 2 THEN v# = v# + ((v# - r#) * 1) + 3
way# = v#
DO
m# = (v# / 3) - INT(v# / 3)
m# = m# * 10
m# = INT(m#)
test1 = (z = 0) AND (m# = 6)
test2 = (z = 1) AND (m# = 0)
test3 = (z = 2) AND (m# = 3)
IF test1 = -1 THEN EXIT DO
IF test2 = -1 THEN EXIT DO
IF test3 = -1 THEN EXIT DO
v# = v# + 1
a$ = INKEY$
IF a$ = "s" THEN STOP
LOOP
way# = v# - way#
PRINT , , z; m#; v#; r#; v# - r#; way#
a$ = INKEY$
INPUT a$
IF a$ = "s" THEN STOP
s1# = r#
IF r# + 94 <= v# THEN r# = r# + 96
IF r# <> s1# THEN flag = 1 ELSE flag = 0
```

```
LOOP
```

```
"C:\Users\Reactor1967\vcns\work\082903B.BAS"
DECLARE SUB dtest2 (v#, r#, way2#)
DECLARE SUB mtest (v#, m#, z!)
DECLARE SUB dtest (v#, r#, way#, z!)
REM it seems I could get a decode here if I could get my equations right
REM for finding my previous r#
redo:
v# = 11
r# = 1
z = 2
flag = 0
RANDOMIZE TIMER
CLS
car = 0
ex = INT(RND * 1000) + 10
DO
car = car + 1
z = INT(RND * 2) + 0
IF flag = 1 THEN z = 2
REM -----------------------------------------------------------
```

```
way# = v#
IF z = 0 THEN v# = v# + ((v# - r#) * 2) + 1
IF z = 1 THEN v# = v# + ((v# - r#) * 2) + 2
IF z = 2 THEN v# = v# + ((v# - r#) * 2) + 3
REM ------------------------------------------------------------
CALL mtest(v#, m#, z)
PRINT z; m#; v#; r#; v# - r#; car; way#
CALL dtest(v#, r#, way#, z)
a$ = INKEY$
REM INPUT a$
IF a$ = "s" THEN STOP
IF a$ = "d" THEN GOTO decode:
REM IF car >= ex THEN GOSUB decode: rem testing decode code.
REM ------------------------------------------------------------
way2# = v#
d# = INT(v# / 2)
s# = (v# / 3) - INT(v# / 3): s# = s# * 10: s# = INT(s#)
dist# = (v# - d#)
dist# = INT(dist# / 2)
dist# = dist# * 2
v# = v# - dist#
s1# = (v# / 3) - INT(v# / 3): s1# = s1# * 10: s1# = INT(s#)
IF s# <> s1# THEN STOP
test = (r# + 96) <= (v# / 2)
```

```
IF test = -1 THEN r# = r# + 96
IF test = -1 THEN flag = 1 ELSE flag = 0
IF flag = 1 THEN z = 2
CALL dtest2(v#, r#, way2#)
LOOP
decode:
store# = v#
m# = (v# / 3) - INT(v# / 3)
m# = m# * 10
m# = INT(m#)
IF m# = 6 THEN z = 0
IF m# = 0 THEN z = 1
IF m# = 3 THEN z = 2
dist# = (v# - r#)
IF z = 0 THEN dist# = dist# - 1
IF z = 1 THEN dist# = dist# - 2
IF z = 2 THEN dist# = dist# - 3
dist# = dist# / 3
dist# = dist# * 2
IF z = 0 THEN dist# = dist# + 1
IF z = 1 THEN dist# = dist# + 2
IF z = 2 THEN dist# = dist# + 3
v# = v# - dist#
IF v# <> way# THEN STOP
```

```
v# = store#
RETURN
IF z = 2 THEN r# = r# - 96

SUB dtest (v#, r#, way#, z)
dist# = (v# - r#)
IF z = 0 THEN dist# = dist# - 1
IF z = 1 THEN dist# = dist# - 2
IF z = 2 THEN dist# = dist# - 3
dist# = dist# / 3
dist# = dist# * 2
IF z = 0 THEN dist# = dist# + 1
IF z = 1 THEN dist# = dist# + 2
IF z = 2 THEN dist# = dist# + 3
IF (v# - dist#) < way# THEN STOP
IF (v# - dist#) > way# THEN STOP
REM PRINT way#; (v# - dist#);
END SUB

SUB dtest2 (v#, r#, way2#)
d# = INT(v# * 2)
dist# = (d# - v#)
REM dist# = INT(dist# / 2)
REM dist# = dist# * 2
```

```
t2# = v# + dist#

PRINT way2#; t2#

IF t2# <> way2# THEN STOP

END SUB


SUB mtest (v#, m#, z)

m# = (v# / 3) - INT(v# / 3)

m# = m# * 10

m# = INT(m#)

IF z = 0 AND m# <> 6 THEN STOP

IF z = 1 AND m# <> 0 THEN STOP

IF z = 2 AND m# <> 3 THEN STOP

END SUB
```

"C:\Users\Reactor1967\vcns\work\082903A.BAS"

```
REM this program takes a number and reduces by what ever you need it reduced

REM to 1/3rd 1/2halve 1/4th ect... It can also reflate the number back.

REM This I thought worked but as of now does not seem to work.

RANDOMIZE TIMER

CLS

DO
```

```
v# = INT(RND * 1999) + 100

d# = INT(v# / 3)

y# = v#

DO

y# = y# - 3

a$ = INKEY$

IF a$ = "s" THEN STOP

LOOP UNTIL (y# - 3) < d#

z# = (y# * 3)

DO

y# = y# + 3

a$ = INKEY$

IF a$ = "s" THEN STOP

LOOP UNTIL (y# + 3) > z#

m1# = (y# / 3) - INT(y# / 3): m1# = m1# * 10: m1# = INT(m1#)

m2# = (v# / 3) - INT(v# / 3): m2# = m2# * 10: m2# = INT(m2#)

test = (y# = v#) AND (m1# = m2#)

PRINT , , v#; d#

REM IF test = -1 THEN PRINT "Tested postive for match"

REM IF test = 0 THEN PRINT "Tested Negative for match"

IF test = 0 THEN STOP

a$ = INKEY$

REM INPUT a$

IF a$ = "s" THEN STOP
```

```
LOOP
```

```
"C:\Users\Reactor1967\vcns\work\082503A.BAS"

REM Maybe create a chart or math equation that could be used to calculate

REM the speed of r#. Every vector would have a specific number that when

REM divied into that vector would give a speed of r# for the next incode.

REM another long shot. Or code a 3(Note this is base two here.)

REM in base then when the vc# number inceases

REM the vc# number being what is divied into the vector v# to give the

REM speed of r# before the next incode.

REM this would have to be done lineraly as to increase or decrease (v# - r#).

v# = 1

r# = 1

RANDOMIZE TIMER

CLS

sr# = 1

DO

z = INT(RND * 2) + 0

REM dist# = (d# * 2) + z + 1

REM dist# = dist# - 2
```

```
r# = r# + INT(dist# / 2)

sv# = v#

v# = v# + (v# - r#) + z + 1

i# = r# - sr#

IF i# = 0 THEN i# = 1

PRINT z; v#; r#; v# - r#; r# - sr#; INT(sv# / i#)

sr# = r#

a$ = INKEY$

REM INPUT a$

IF a$ = "s" THEN STOP

r# = r# + INT((v# - r#) / 2)

d# = v# - r#

LOOP
```

"C:\Users\Reactor1967\vcns\work\082403A.BAS"

```
REM Getting better. WE know have a third variable I call vc#. This variable

REM is what you divided v# by to get the speed of r# before the next code.

REM Nothing as change you still have to know when to increment and decrement

REM which has been a problem before but if I can use base three and 1 value
```

```
REM in base three to tell the program when to increment and decrement then

REM the next two values can be used as binary to code binary.

v# = 1

r# = 1

RANDOMIZE TIMER

CLS

DO

z = INT(RND * 3) + 0

sv# = v#

v# = v# + ((v# - r#) * 2) + z + 1

IF r# - sr# = 0 THEN GOTO skip:

m# = (v# / 3) - INT(v# / 3)

m# = m# * 10

m# = INT(m#)

PRINT z; m#; v#; r#; v# - r#; r# - sr#; sv# / (r# - sr#)

skip:

sr# = r#

a$ = INKEY$

IF a$ = "s" THEN STOP

r# = r# + (v# - sv#) - z - 1

r# = INT(r# / 3)

r# = r# * 3

LOOP
```

```
"C:\Users\Reactor1967\vcns\work\082303D.BAS"

REM I always seem to come back to the speed of r#

REM here the speed of r# is 1.5 divided into the distance between v# - r#

REM now if I can do that before decoding I can lick this. Anyway or I can

REM code a binary 3 every time the speed of r# increments and a binary 0

REM and 1 to code data.

v# = 8

r# = 1

CLS

RANDOMIZE TIMER

DO

z = INT(RND * 3) + 0

sr# = INT((v# - r#) / 1.5)

sr# = INT(sr# / 3)

sr# = sr# * 3

r# = r# + sr#

IF z = 0 THEN v# = v# + ((v# - r#) * 2) + 1

IF z = 1 THEN v# = v# + ((v# - r#) * 2) + 2

IF z = 2 THEN v# = v# + ((v# - r#) * 2) + 3
```

```
m# = (v# / 3) - INT(v# / 3)

m# = m# * 10

m# = INT(m#)

PRINT z; m#; v#; r#; v# - r#; sr#

a$ = INKEY$

REM INPUT a$

IF a$ = "s" THEN STOP

LOOP
```

"C:\Users\Reactor1967\vcns\work\082303C.BAS"

```
REM trying to find a way to calculate the speed of r# and keeping up

REM with it mathmaticaly so I can decode.

REM 1. code your data

REM 2. Figure you speed of v#

REM 3. calculate your speed of r#

REM 4. decode

REM 5. figure your speed of v#

REM 6. figure your speed of r#

v# = 4

r# = 1

l = 0

b = 0
```

```
CLS
RANDOMIZE TIMER
DO
z = INT(RND * 3) + 0
sv# = (v# - r#) * 2
IF z = 0 THEN v# = v# + ((v# - r#) * 2) + 1
IF z = 1 THEN v# = v# + ((v# - r#) * 2) + 2
IF z = 2 THEN v# = v# + ((v# - r#) * 2) + 3
m# = (v# / 3) - INT(v# / 3)
m# = m# * 10
m# = INT(m#)
PRINT z; m#; v#; r#; v# - r#; y#
y# = INT(sv# / 3)
y# = y# * 3
r# = r# + y#
a$ = INKEY$
IF a$ = "s" THEN STOP
LOOP
```

```
"C:\Users\Reactor1967\vcns\work\082303B.BAS"
REM trying to find a way to calculate the speed of r# and keeping up
REM with it mathmaticaly so I can decode.
```

```
REM 1. code your data

REM 2. Figure you speed of v#

REM 3. calculate your speed of r#

REM 4. decode

REM 5. figure your speed of v#

REM 6. figure your speed of r#

v# = 4

r# = 1

l = 0

b = 0

CLS

RANDOMIZE TIMER

DO

z = INT(RND * 3) + 0

sv# = v# - r#

IF z = 0 THEN v# = v# + ((v# - r#) * 2) + 1

IF z = 1 THEN v# = v# + ((v# - r#) * 2) + 2

IF z = 2 THEN v# = v# + ((v# - r#) * 2) + 3

m# = (v# / 3) - INT(v# / 3)

m# = m# * 10

m# = INT(m#)

REM PRINT z; m#; v#; r#; v# - r#; (v# - r#) - x3#; v# - x1#; r# - x2#;
sv#; y#

REM PRINT z; m#; v#; r#; v# - r#; v# - x1#; r# - x2#; sv#; y#
```

```
PRINT , , z; m#; v#; r#; v# - r#; y#; (v# - r#) - sv# - z - 1; v# - x1#

x1# = v#

x2# = r#

x3# = (v# - r#)

a$ = INKEY$

INPUT a$

IF a$ = "s" THEN STOP

REM d# = (v# - r#) before coding

y# = (v# - r#) - sv# - z - 1 + l - b: REM this equals ((((v# - r#) - n) / base)
* 2)

y# = INT(y# / 3)

y# = y# * 3

r# = r# + y#


LOOP




"C:\Users\Reactor1967\vcns\work\082303A.BAS"

REM it seems like this could of worked

REM when you decode and v# - r# is < x# then

REM thats when you take x# down a notch

REM and subtract it from r# to get your

REM new r#. This could be self coding but
```

```
REM it is still just a good trick nothing else.

v# = 7

r# = 1

x# = 6

RANDOMIZE TIMER

CLS

DO

z = INT(RND * 2) + 0

IF z = 0 THEN v# = v# + (v# - r# + 1)

IF z = 1 THEN v# = v# + (v# - r# + 2)

m# = (v# / 4) - INT(v# / 4)

m# = m# * 10

m# = INT(m#)

PRINT z; m#; v#; r#; v# - r#; x#; v# - sv#; r# - sr#

sr# = r#

sv# = v#

a$ = INKEY$

REM INPUT a$

IF a$ = "s" THEN STOP

r# = r# + x#

IF (v# - r#) - x# >= 1 THEN x# = x# + 2

LOOP
```

```
"C:\Users\Reactor1967\vcns\work\082203A.BAS"
REM It seems the only way to create a self coding program it
REM base 3 that uses 1 & 2 to control the system and 3 to
REM tell when to stop is to somehow create two states with
REM v# - r# or d# were N tells the previous state.
REM the above program was copied over.
REM revized version here.
REM So what you do to v# you must do to r#. In math that is as
REM old as the hills. Here decode to find your previous v#
REM after that observe N that tells you what to subtract from
REM v# then subtract that from r# minus another 4 to get your
REM previous r#. This could really work if you can find your N.
REM problem here is No sysmetery.
CLS
v# = 5
r# = 1
v2# = 1
RANDOMIZE TIMER
x# = 0
DO
z = INT(RND * 2) + 0
sv# = v#
IF z = 0 THEN v# = v# + (v# - r#)
```

```
IF z = 1 THEN v# = v# + (v# - r# + 2)

m# = (v# / 4) - INT(v# / 4)

m# = m# * 10

m# = INT(m#)

PRINT , , z; m#; v#; r#; v# - r#; r# / 4

r# = r# + (v# - sv#)

INPUT a$

IF a$ = "s" THEN STOP

LOOP
```

```
"C:\Users\Reactor1967\vcns\work\082103A.BAS"

REM this program proves there may not be a need for r#

REM its possible to use fractions to get what we need

REM instead of a reference point. The reference point helps

REM serve the same need as fractions.

v# = 879345

r# = 879345

RANDOMIZE TIMER

CLS

REM GOTO prog4:

DO

REM v# - (v# - r#) * 2 is the same as v# - ((r# - v#) * 2)
```

```
z = INT(RND * 3) + 0

a# = v#

b# = v# - ((r# - v#) * 2)

IF z = 0 THEN v# = v# - ((r# - v#) * 2) - 1

IF z = 1 THEN v# = v# - ((r# - v#) * 2) - 2

IF z = 2 THEN v# = v# - ((r# - v#) * 2) - 3

m# = (v# / 3) - INT(v# / 3)

m# = m# * 10

m# = INT(m#)

PRINT z; m#; v#; r#; r# - v#, a#; b#

INPUT a$

IF a$ = "s" THEN STOP

LOOP

prog2:

DO

v# = INT(RND * 1000) + 1

v# = (v# * 3) + 1

sv# = v#

z = INT(RND * 3) + 0

IF z = 0 THEN v# = v# + 1

IF z = 1 THEN v# = v# + 2

IF z = 2 THEN v# = v# + 3

m# = (v# / 3) - INT(v# / 3)

m# = m# * 10
```

```
m# = INT(m#)

PRINT , , z; m#; v#; sv# / 3

INPUT a$

IF a$ = "s" THEN STOP

LOOP

prog3:

v# = 1

r# = 1

CLS

redo:

z = 2

DO

z = INT(RND * 2) + 0

IF z = 0 THEN v# = v# + ((v# - r#) * 2) + 1

IF z = 1 THEN v# = v# + ((v# - r#) * 2) + 2

IF z = 2 THEN v# = v# + ((v# - r#) * 2) + 3

m# = (v# / 3) - INT(v# / 3)

m# = m# * 10

m# = INT(m#)

PRINT , , z; m#; v#; r#; v# - r#

INPUT a$

IF a$ = "s" THEN STOP

LOOP UNTIL v# - r# >= 100

prog4:
```

```
r# = 1001

v# = 1001

CLS

DO

z = INT(RND * 3) + 0

dist# = v# - r#

dist# = dist# * 2

v# = v# - dist#

IF z = 0 THEN v# = v# - 1

IF z = 1 THEN v# = v# - 2

IF z = 2 THEN v# = v# - 3

m# = (v# / 3) - INT(v# / 3)

m# = m# * 10

m# = INT(m#)

PRINT , , z; m#; v#; r#

INPUT a$

IF a$ = "s" THEN STOP

LOOP
```

```
"C:\Users\Reactor1967\vcns\work\082003T2.BAS"

REM the program on the right is the control program.

REM the program on the left is the experiment program.
```

```
REM the experiment program calculates what its r's should be
REM and the control program matches as close as possible those
REM r's for both coding and decoding.
v2# = 100
r# = 1
v# = 2
sv# = 1
z = 0
lb# = 0
CLS
RANDOMIZE TIMER
DO
z2 = INT(RND * 3) + 0
IF z2 = 0 THEN v2# = v2# + ((v2# - r#) * 2) + 1
IF z2 = 1 THEN v2# = v2# + ((v2# - r#) * 2) + 2
IF z2 = 2 THEN v2# = v2# + ((v2# - r#) * 2) + 3
m# = (v2# / 3) - INT(v2# / 3)
m# = m# * 10
m# = INT(m#)
PRINT z2; m#; v2#; r#; v2# - r#; r# - sr#; ; v#; r#; v# - r#; car; car / sv#
sv# = v# - r#
sr# = r#
IF v2# - r# < 100 THEN lb# = 0
IF v2# - r# > 10000 THEN lb# = 4
```

```
a$ = INKEY$

REM INPUT a$

IF a$ = "s" THEN STOP

a# = INT((v2# - r#) / 3)

car = 0

DO

car = car + 1

store1# = v#

store2# = r#

test = (v# >= (r# + 5))

IF test = -1 THEN r# = r# + 6

IF test = -1 THEN z = 1 ELSE z = 0

IF z = 0 THEN v# = v# + (v# - r# + 1)

IF z = 1 THEN v# = v# + (v# - r# + 2)

REM PRINT , , z; v#; r#; v# - r#; (r# / 3) - INT(r# / 3)

IF r# > v# THEN STOP

REM a$ = INKEY$

REM INPUT a$

REM IF a$ = "s" THEN STOP

REM IF a$ = "d" THEN EXIT DO

REM lb# = 4: REM lb# = 0 when distance gets too low and lb# = 4 when
distance getts

REM         too high.

IF (v2# - r#) < (a# - lb#) THEN EXIT DO
```

```
LOOP
v# = store1#
r# = store2#
LOOP
```

```
"C:\Users\Reactor1967\vcns\work\082003T1.BAS"
REM this program is a template for a coding program using base 3
REM as it codes the distance between v# and r# will be divided by
REM 3 to get a new distance for coding. This program will try to fit
REM that distance as closely as possible. When decoding that same
REM distance will be multiplied by 3 to get the previous distance.
REM again this program which can code and decode will try to bring r#
REM as close as possible to that distance but it will have to be exact
REM on the r's for coding and decoding. If I can match the distances
REM perfectly then I can get a data coding program instead of a self
REM coding and decoding progam. This program performs some good
math
REM tricks but it needs to be able to store information mathmaticly.
v# = 1
r# = 1
z = 0
CLS
```

```
RANDOMIZE TIMER
DO
IF z = 0 THEN v# = v# + (v# - r# + 1)
IF z = 1 THEN v# = v# + (v# - r# + 2)
m# = (v# / 3) - INT(v# / 3)
m# = m# * 10
m# = INT(m#)
PRINT , , z; m#; v#; r#; v# - r#; (r# / 3) - INT(r# / 3)
IF r# > v# THEN STOP
a$ = INKEY$
REM INPUT a$
IF a$ = "s" THEN STOP
IF a$ = "d" THEN EXIT DO
test = (v# >= (r# + 5))
IF test = -1 THEN r# = r# + 6
IF test = -1 THEN z = 1 ELSE z = 0
LOOP
DO
IF (v# / 2) - INT(v# / 2) = 0 THEN z = 0
IF (v# / 2) - INT(v# / 2) = .5 THEN z = 1
PRINT , , z; v#; r#; v# - r#; (r# / 3) - INT(r# / 3)
IF v# <= 1 THEN SYSTEM
a$ = INKEY$
REM INPUT a$
```

```
IF a$ = "s" THEN STOP
dist# = (v# - r#)
IF (dist# / 2) - INT(dist# / 2) = .5 THEN dist# = dist# - 1
dist# = dist# / 2
dist# = dist# + 1
v# = v# - dist#
IF z = 1 THEN r# = r# - 6
LOOP
```

```
"C:\Users\Reactor1967\vcns\work\082003I.BAS"
v# = 6
r# = 1
CLS
RANDOMIZE TIMER
DO
z = INT(RND * 3) + 0
IF z = 0 THEN v# = v# + ((v# - r#) * 2) + 1
IF z = 1 THEN v# = v# + ((v# - r#) * 2) + 2
IF z = 2 THEN v# = v# + ((v# - r#) * 2) + 3
m# = (v# / 3) - INT(v# / 3)
m# = m# * 10
m# = INT(m#)
```

```
PRINT , , z; m#; v#; r#; v# - r#
a$ = INKEY$
REM INPUT a$
IF a$ = "s" THEN STOP
dist# = v# - r#
m# = (dist# / 3) - INT(dist# / 3)
m# = m# * 10
m# = INT(m#)
x# = INT(((v# - r#) / 3))
REM IF v# - r# < 10 THEN GOTO skip:
DO
y# = (x# / 3) - INT(x# / 3)
y# = y# * 10
y# = INT(y#)
test = (y# = m#)
IF test = -1 THEN EXIT DO
x# = x# + 1
REM x# = x# - 1
a$ = INKEY$
IF a$ = "s" THEN STOP
LOOP
r# = v# - x#
REM skip:
LOOP
```

```
"C:\Users\Reactor1967\vcns\work\082003H.BAS"
REM Everything works by fractions and geometery and with a reference
REM point. You have to trianglelate your position of what ever you want.
REM try a self decoding number that always codes with the remainder
REM being the same when devided by one number but different when
REM devided by another. You will have to be able to decode it.
REM then try fitting to the picture you draw by taking your
REM fractions. It should fit as closely as possible to get your
REM answer for the next decode.
v# = 1
r# = 1
CLS
RANDOMIZE TIMER
DO
z = INT(RND * 3) + 0
IF z = 0 THEN v# = v# + ((v# - r#) * 2) + 1
IF z = 1 THEN v# = v# + ((v# - r#) * 2) + 2
IF z = 2 THEN v# = v# + ((v# - r#) * 2) + 3
m# = (v# / 3) - INT(v# / 3)
m# = m# * 10
m# = INT(m#)
```

```
PRINT , , z; m#; v#; r#; v# - r#
a$ = INKEY$
REM INPUT a$
IF a$ = "s" THEN STOP
dist# = v# - r#
dist# = INT(dist# / 3)
r# = v# - dist#
r# = INT(r# / 3)
r# = r# * 3
LOOP
```

```
"C:\Users\Reactor1967\vcns\work\082003G.BAS"
REM (13874 and 4624) are you reference points. So again you have to
REM know your reference points and change them around. If you know
REM when to change your reference points and what to change them
REM to you can code data from here to enterenity and decode it to.
v# = 13874
c# = 4624
v1# = v#
CLS
REM GOTO prog2:
```

```
DO
m# = (c# / 3) - INT(c# / 3)
m# = m# * 10
m# = INT(m#)
PRINT , , v#; c#; m#, "13874 v#"; "4624 r#"
INPUT a$
IF a$ = "s" THEN STOP
v# = v# - 1
c# = c# - 3
LOOP
STOP
prog2:
RANDOMIZE TIMER
DO
z = INT(RND * 3) + 1
x# = ((ABS(v1# - v#)) * 3) + c#
v1# = x#
IF z = 0 THEN v1# = v1# + 1
IF z = 1 THEN v1# = v1# + 2
IF z = 2 THEN v1# = v1# + 3
m# = (v1# / 3) - INT(v1# / 3)
m# = m# * 10
m# = INT(m#)
PRINT , , z; m#; v1#;
```

```
INPUT a$

IF a$ = "s" THEN STOP

LOOP
```

"C:\Users\Reactor1967\vcns\work\082003F.BAS"

```
REM again this proves that you can only do with a reference point

REM that you can do by multiplying. You can multiply any number

REM by 2 and and 1 then add it to your original number and divide

REM it by 3 to get .33333. But when you look at both numbers one

REM of those numbers serves as v# and the other serves as the reference

REM point. You can do the same thing going down but you need a
reference

REM point.

v# = 187352

CLS

DO

test# = INT(v# / 3)

DO

m# = test#

m# = (m# / 3) - INT(m# / 3)

m# = m# * 10

m# = INT(m#)
```

```
IF m# = 3 THEN EXIT DO

test# = test# - 1

a$ = INKEY$

IF a$ = "s" THEN STOP

LOOP

PRINT , , v# + (v# * 2) + 1; v#; test#

a$ = INKEY$

IF v# < 100 THEN INPUT a$

IF a$ = "s" THEN STOP

v# = v# + 1

LOOP
```

"C:\Users\Reactor1967\vcns\work\082003E.BAS"

```
REM again this proves that you can only do with a reference point

REM that you can do by multiplying. You can multiply any number

REM by 2 and and 1 then add it to your original number and divide

REM it by 3 to get .33333. But when you look at both numbers one

REM of those numbers serves as v# and the other serves as the reference

REM point. You can do the same thing going down but you need a
reference

REM point.

v# = 187352
```

```
CLS

DO

test# = INT(v# / 3)

DO

m# = test#

m# = (m# / 3) - INT(m# / 3)

m# = m# * 10

m# = INT(m#)

IF m# = 3 THEN EXIT DO

test# = test# - 1

a$ = INKEY$

IF a$ = "s" THEN STOP

LOOP

PRINT , , v# + (v# * 2) + 1; v#; test#

a$ = INKEY$

IF v# < 100 THEN INPUT a$

IF a$ = "s" THEN STOP

v# = v# + 1

LOOP
```

"C:\Users\Reactor1967\vcns\work\082003E (2).BAS"

```
REM Here are a couple of equations for doing this stuff.
```

```
REM now if I just can use multiples of v# and be able to decode
REM with it I might have something. In short use factors
REM Nothing meaniful here this was a canvas I did not finish painting on.
REM b# = (v# * 2) + 1
REM  v# = (b# - 1) / 3
v# = 9898787
CLS
RANDOMIZE TIMER
DO
z = INT(RND * 3) + 0
x# = INT(v# / 3)
v# = x# + 50
DO
v# = v# - 1
m# = (v# / 3) - INT(v# / 3)
m# = m# * 10
m# = INT(m#)
test = (m# = 3) AND (v# <= x#)
a$ = INKEY$
IF a$ = "s" THEN STOP
IF test = -1 THEN EXIT DO
LOOP
IF z = 0 THEN v# = v# - 1
```

```
IF z = 1 THEN v# = v# - 2

IF z = 2 THEN v# = v# - 3

m# = v#

m# = (m# / 3) - INT(m# / 3)

m# = m# * 10

m# = INT(m#)

PRINT , , z; m#; v#

sv# = v#

INPUT a$

IF a$ = "s" THEN STOP

LOOP
```

```
"C:\Users\Reactor1967\vcns\work\082003D.BAS"

REM Here are a couple of equations for doing this stuff.

REM now if I just can use multiples of v# and be able to decode

REM with it I might have something. In short use factors

REM b# = (v# * 2) + 1

REM  v# = (b# - 1) / 3

v# = 1

CLS
```

```
RANDOMIZE TIMER

DO

z = INT(RND * 3) + 0

b# = (v# * 2) + 1

REM b# = (v# / 2) + 1

REM IF b# - INT(b#) = .5 THEN b# = b# - .5

IF z = 0 THEN v# = v# + 1 + b#

IF z = 1 THEN v# = v# + 2 + b#

IF z = 2 THEN v# = v# + 3 + b#

m# = v#

m# = (m# / 3) - INT(m# / 3)

m# = m# * 10

m# = INT(m#)

PRINT , , z; m#; v#

INPUT a$

IF a$ = "s" THEN STOP

IF a$ = "d" THEN EXIT DO

LOOP
```

```
"C:\Users\Reactor1967\vcns\work\082003C.BAS"

REM b# = (v# * 2) + 1
```

```
REM v# = ((v# + b#) - 1) / 3

v# = 1

b# = 1

CLS

DO

redo:

m# = v# + b#

m# = (m# / 3) - INT(m# / 3)

m# = m# * 10

m# = INT(m#)

IF m# <> 3 THEN b# = b# + 1

a$ = INKEY$

IF a$ = "s" THEN STOP

IF m# <> 3 THEN GOTO redo:

PRINT , , m#; v#; b#; v# + b#

INPUT a$

IF a$ = "s" THEN STOP

v# = v# + 1

LOOP
```

"C:\Users\Reactor1967\vcns\work\082003B.BAS"

REM We know how to keep the sysmetery with r# which is r# = r# + base or

REM by multiples of that. But how do we keep the sysmetery with v# which

REM is v# = v# + ((v# - r#) * base - 1). Here without N v# / base is always

REM the same thing. Well thats because above v# + M# without N is always

REM the same thing. By keeping r# and v# aligned when we add N we have

REM have sysmetery which is we can divide v# by the base and find N

REM by looking at the remainder. Knowing this allows us to change the

REM equations around Now and experiment.

```
v# = 1

b1# = 1: b2# = 2: b3# = 3

CLS

RANDOMIZE TIMER

DO

a# = v# + b1#

b# = v# + b2#

c# = v# + b3#

a# = (a# / 3) - INT(a# / 3)

a# = a# * 10

a# = INT(a#)

b# = (b# / 3) - INT(b# / 3)

b# = b# * 10
```

```
b# = INT(b#)

c# = (c# / 3) - INT(c# / 3)

c# = c# * 10

c# = INT(c#)

test1 = (a# = 3)

test2 = (b# = 3)

test3 = (c# = 3)

IF test1 = -1 THEN v# = v# + b1#

IF test2 = -1 THEN v# = v# + b2#

IF test3 = -1 THEN v# = v# + b3#

z = INT(RND * 3) + 0

IF z = 0 THEN v# = v# + 1

IF z = 1 THEN v# = v# + 2

IF z = 2 THEN v# = v# + 3

m# = (v# / 3) - INT(v# / 3)

m# = m# * 10

m# = INT(m#)

PRINT , , z; m#; v#; b1#; b2#; b3#; test1; test2; test3

INPUT a$

IF a$ = "s" THEN STOP

b1# = b1# + 3

b2# = b2# + 3

b3# = b3# + 3

LOOP
```

"C:\Users\Reactor1967\vcns\work\082003A.BAS"

```
REM We know how to keep the sysmetery with r# which is r# = r# + base
or
REM by multiples of that. But how do we keep the sysmetery with v#
which
REM is v# = v# + ((v# - r#) * base - 1). Here without N v# / base is
always
REM the same thing. Well thats because above v# + M# without N is
always
REM the same thing. By keeping r# and v# aligned when we add N we
have
REM have sysmetery which is we can divide v# by the base and find N
REM by looking at the remainder. Knowing this allows us to change the
REM equations around Now and experiment.
v# = 1
b# = 0
CLS
DO
sb# = b#
DO
b# = b# + 1
m# = (v# + b#)
m# = m# / 3
```

```
m# = m# - INT(m#)

m# = m# * 10

m# = INT(m#)

test = (m# = 3)

IF test = -1 THEN EXIT DO

a$ = INKEY$

IF a$ = "s" THEN STOP

LOOP

z = INT(RND * 3) + 0

IF z = 0 THEN v# = v# + b# + 1

IF z = 1 THEN v# = v# + b# + 2

IF z = 2 THEN v# = v# + b# + 3

y# = INT(((v# / 3) - INT(v# / 3)) * 10)

PRINT , , z; y#; v#; b#

a$ = INKEY$

INPUT a$

IF a$ = "s" THEN STOP

LOOP
```

```
"C:\Users\Reactor1967\vcns\work\081803A.BAS"

REM This program shows that you can use d# to determine N but it does
```

```
REM no good because you have to know N to determine D. But its
interesting

REM though.

REM Im thinking though use the program to calculate distances and

REM speed of r#. Maybe something can be written in stone here to help

REM code and decode.

RANDOMIZE TIMER

CLS

redo:

dist# = 10

DO

x# = INT(dist# / 2)

b# = INT(x# / 4)

b# = b# * 4

a# = dist# - b#

m# = (a# / 4) - INT(a# / 4)

m1# = (dist# / 4) - INT(dist# / 4)

PRINT , , z; m1#; dist#; a#; b#; m#

a$ = INKEY$

REM INPUT a$

IF a$ = "s" THEN STOP

z = INT(RND * 2) + 0

IF z = 0 THEN dist# = (a# * 2) + 1

IF z = 1 THEN dist# = (a# * 2) + 2
```

```
LOOP

"C:\Users\Reactor1967\vcns\work\081703A.BAS"
CLS
RANDOMIZE TIMER
v# = 2
r# = 1
d# = 1
z = 1
RANDOMIZE TIMER
CLS
DO
REM z = INT(RND * 2) + 0
REM IF z = 1 THEN z = 0 ELSE z = 1
z = 0
d# = v# - r#
IF z = 0 THEN v# = v# + (v# - r# + 1)
IF z = 1 THEN v# = v# + (v# - r# + 2)
PRINT , z; v#; r#; v# - r#; (v# - r#) - vr#; d#
vr# = v# - r#
a$ = INKEY$
REM INPUT a$
```

```
IF a$ = "s" THEN STOP
IF a$ = "d" THEN EXIT DO
d# = d# + 2
REM IF z = 0 THEN d# = d# + 2 ELSE d = d# + 3
r# = v# - d#
LOOP
DO
PRINT , z; v#; r#; v# - r#
IF v# = 4 THEN EXIT DO
a$ = INKEY$
REM INPUT a$
IF a$ = "s" THEN STOP
sv# = v# - r#
dist# = v# - r#
IF (dist# / 2) - INT(dist# / 2) = .5 THEN dist# = dist# - 1
dist# = dist# / 2
dist# = dist# + 1
v# = v# - dist#
sv# = sv# - 4
r# = v# - sv#
LOOP
```

```
"C:\Users\Reactor1967\vcns\work\081603F.BAS"
REM the speed of int((v# - N) / 4) * 4 = speed of r#
REM I wondered how to take care of the little rounding errors.
REM this seems to be it. The interger of the speed of v# minus
REM N divided by some number times some number equals the speed
REM of r#. How ever your rounding error is just increase the number
REM you divide by. Also to keep true sysmetery that number should
REM be the (base * 2) or some multiple of that.
REM you may have to use all even or odd distances no bother though
REM you can use 2 or 4 for N or 1 or 3 for N or various combinations
REM of that as long as there is sysmetery in your vectors for finding
REM N.
CLS
RANDOMIZE TIMER
DO
redo:
vr# = INT(RND * 100) + 1
test = (vr# / 2) - INT(vr# / 2)
a$ = INKEY$
IF a$ = "s" THEN STOP
IF test = .5 THEN GOTO redo:
d# = 1
DO
y# = vr# - d#
```

```
test = INT(y# / 4) = INT(d# / 4)
IF test = -1 THEN EXIT DO
d# = d# + 1
IF d# > vr# THEN STOP
a$ = INKEY$
IF a$ = "s" THEN STOP
LOOP
PRINT , , vr#; y#; d#; test
a$ = INKEY$
REM INPUT a$
IF a$ = "s" THEN STOP
LOOP
```

```
"C:\Users\Reactor1967\vcns\work\081603E.BAS"
v# = 101
r# = 51
CLS
RANDOMIZE TIMER
REM the speed of r# = (((v# - r#) - N) / 2) - (N - 1)
REM Divide v# / 4 and look at remainder for N
REM a binary 0 = .25
REM a binary 1 = .75
```

```
REM if v# = a binary 0 then N = 2

REM if v# = a binary 1 then N = 4

DO

z = INT(RND * 2) + 0

d# = 1

DO

IF z = 0 THEN x# = (d# * 2) + 2

IF z = 1 THEN x# = (d# * 2) + 4

y# = INT(x# / 2)

stable# = ((v# - r#) - d#)

test = ((v# - r#) - d# <= y#) AND ((stable# / 4) - INT(stable# / 4) = 0)

IF test = -1 THEN EXIT DO

d# = d# + 1

LOOP

REM IF ((v# - r#) - d# < y#) THEN d# = d# - 4

r# = v# - d#

IF z = 0 THEN v# = v# + (v# - r# + 2)

IF z = 1 THEN v# = v# + (v# - r# + 4)

m# = (v# / 4) - INT(v# / 4)

PRINT , z; m#; v#; r#; v# - r#; r# - sr#; (v# - r#) / 2; ((v# - r#) / 4) -
INT((v# - r#) / 4)

sr# = r#

a$ = INKEY$

IF a$ = "s" THEN STOP
```

```
IF a$ = "d" THEN EXIT DO

LOOP

STOP

REM DO

test = (v# / 4) - INT(v# / 4)

IF test = .25 THEN z = 0

IF test = .75 THEN z = 1

y# = v# - r#

dist# = v# - r#

dist# = dist# / 2

IF z = 0 THEN dist# = dist# + 1

IF z = 1 THEN dist# = dist# + 2

v# = v# - dist#

dist# = y#

IF z = 0 THEN dist# = dist# - 2

IF z = 1 THEN dist# = dist# - 4

dist# = dist# / 2

IF z = 0 THEN dist# = dist# - 1

IF z = 1 THEN dist# = dist# - 3
```

```
"C:\Users\Reactor1967\vcns\work\081603D.BAS"

REM This program proves that you can make the speed of r# the function
```

```
REM of the distance between v# and R#

RANDOMIZE TIMER

CLS

DO

DO

vr# = INT(RND * 100) + 1

IF (vr# / 2) - INT(vr# / 2) = 0 THEN EXIT DO

a$ = INKEY$

IF a$ = "s" THEN STOP

LOOP

d# = 1

z = INT(RND * 2) + 0

DO

IF z = 0 THEN sr# = (d# * 2) + 2

IF z = 1 THEN sr# = (d# * 2) + 4

test# = INT(sr# / 2)

test1# = (vr# - d# <= test#)

REM PRINT , , z; vr#; d#; sr#; vr# - d#; test#

IF test1# = -1 THEN EXIT DO

a$ = INKEY$

REM INPUT a$

IF a$ = "s" THEN STOP

d# = d# + 1

LOOP
```

```
PRINT , , z; vr#; d#; sr#; test#
INPUT a$
IF a$ = "s" THEN STOP
LOOP
```

```
"C:\Users\Reactor1967\vcns\work\081603C.BAS"
RANDOMIZE TIMER
v# = 100
CLS
DO
REM v# = INT(RND * 5000)
x# = INT(v# / 4#)
r# = x# * 4#
DO
IF v# - r# < 30 THEN r# = r# - 4
a$ = INKEY$
IF a$ = "s" THEN STOP
LOOP UNTIL v# - r# >= 30
z = INT(RND * 2) + 0
IF z = 0 THEN v# = v# + (v# - r# + 1)
IF z = 1 THEN v# = v# + (v# - r# + 2)
m# = (v# / 4) - INT(v# / 4)
```

```
PRINT , , z; m#; v#; r#; r# - sr#; v# - r#
sr# = r#
a$ = INKEY$
REM INPUT a$
IF a$ = "s" THEN STOP
LOOP
```

"C:\Users\Reactor1967\vcns\work\081603B.BAS"

```
REM This program proves that you can make the speed of r# the function
REM of the distance between v# and R#
RANDOMIZE TIMER
CLS
DO
DO
vr# = INT(RND * 100) + 1
IF (vr# / 2) - INT(vr# / 2) = 0 THEN EXIT DO
a$ = INKEY$
IF a$ = "s" THEN STOP
LOOP
d# = 1
z = INT(RND * 2) + 0
DO
```

```
IF z = 0 THEN sr# = (d# * 2) + 2

IF z = 1 THEN sr# = (d# * 2) + 4

test# = INT(sr# / 2)

test1# = (vr# - d# <= test#)

REM PRINT , , z; vr#; d#; sr#; vr# - d#; test#

IF test1# = -1 THEN EXIT DO

a$ = INKEY$

REM INPUT a$

IF a$ = "s" THEN STOP

d# = d# + 1

LOOP

PRINT , , z; vr#; d#; sr#; test#

INPUT a$

IF a$ = "s" THEN STOP

LOOP
```

```
"C:\Users\Reactor1967\vcns\work\081603A.BAS"

REM 1 = 0 d# = 45

REM 0 = .66667 d# = 30

REM 2 = .333333 d# = 42

v# = 100

REM r# = 1
```

```
RANDOMIZE TIMER
CLS
DO
z = INT(RND * 3) + 0
IF z = 0 THEN r# = v# - 30
IF z = 1 THEN r# = v# - 45
IF z = 2 THEN r# = v# - 42
IF z = 0 THEN v# = v# + ((v# - r#) * 2) + 1
IF z = 1 THEN v# = v# + ((v# - r#) * 2) + 2
IF z = 2 THEN v# = v# + ((v# - r#) * 2) + 3
m# = (v# / 3) - INT(v# / 3)
PRINT z; v#; r#; v# - r#; r# - sr#
sr# = r#
a$ = INKEY$
REM INPUT a$
IF a$ = "s" THEN STOP
REM test1 = ((v# - r#) >= 96) AND ((v# - r#) < 198)
REM test2 = ((v# - r#) >= 198) AND ((v# - r#) < 300)
REM test3 = ((v# - r#) >= 300)
REM IF test1 = -1 THEN r# = r# + 96
REM IF test2 = -1 THEN r# = r# + 198
REM IF test3 = -1 THEN r# = r# + 300
LOOP
```

```
"C:\Users\Reactor1967\vcns\work\081503A.BAS"
v# = 100
r# = 50
CLS
RANDOMIZE TIMER
DO
z = INT(RND * 2) + 0
d# = 1
DO
IF z = 0 THEN x# = (d# * 2) + 2
IF z = 1 THEN x# = (d# * 2) + 4
y# = INT(x# / 2)
test = ((v# - r#) - d# <= y#)
IF test = -1 THEN EXIT DO
d# = d# + 1
LOOP
REM IF ((v# - r#) - d# < y#) THEN d# = d# - 4
r# = v# - d#
IF z = 0 THEN v# = v# + (v# - r# + 2)
IF z = 1 THEN v# = v# + (v# - r# + 4)
m# = (v# / 4) - INT(v# / 4)
PRINT , , z; m#; v#; r#; v# - r#; r# - sr#; (v# - r#) / 2
```

```
sr# = r#
a$ = INKEY$
IF a$ = "s" THEN STOP
LOOP
```

```
"C:\Users\Reactor1967\vcns\work\081003A.BAS"
z = -1
d# = 2
t$ = "c:\test.txt"
CLS
DO
a# = INT(d# / 2)
REM a# = d#
OPEN t$ FOR APPEND AS #1
d1# = 1
d2# = 2
WRITE #1, d#
FOR count = 1 TO a#
WRITE #1, (d1# / d#) - INT(d1# / d#), (d2# / d#) - INT(d2# / d#)
a$ = INKEY$
IF a$ = "s" THEN CLOSE #1
IF a$ = "s" THEN SYSTEM
```

```
d1# = d1# + 1

d2# = d2# + 2

NEXT count

WRITE #1, z

CLOSE #1

PRINT "Written d = "; d#

a$ = INKEY$

REM INPUT a$

IF a$ = "s" THEN CLOSE #1

IF a$ = "s" THEN SYSTEM

d# = d# + 1

IF d# = 101 THEN EXIT DO

LOOP
```

```
"C:\Users\Reactor1967\vcns\work\080903B.BAS"

REM when converting numbers to a specific number using fractions like

REM Im doing use the low side of a postive number being used for division

REM then take a multiple of that number and use the upper side of that number

REM for fractions. This cover both bases. Try this with both even and odd

REM numbers.
```

```
f# = 2
CLS
DO
d1# = 1
d2# = 2
FOR count = 1 TO INT(f# / 2)
PRINT d1# / f#; d2# / f#, f#
INPUT a$
IF a$ = "s" THEN STOP
d1# = d1# + 1
d2# = d2# + 2
NEXT count
a$ = INKEY$
IF a$ = "s" THEN STOP
f# = f# + 1
LOOP
```

```
"C:\Users\Reactor1967\vcns\work\080903A.BAS"
REM From looking at this program you can not tell how it is useful but
REM I will explain. Well the goal here is to take any number and convert
REM it to a postive number and be able to convert it back to its orginal
```

REM number. What this will be used for is adding N to the positive number

REM then taking that number and converting it to a postive number and add

REM N to that and so on. Then when decoding we will know N so we will

REM subtract N which will give us our postive number which we will convert

REM back to its orginal number then find N there and subtract that which again

REM will give us our postiive number which we will convert and so on to

REM decode.

REM To do this we can use fractions. Divide a number by another number. Look

REM at its fraction.

REM This tells us what fraction we want to goto. Now add or subtract

REM from our orginal number before dividing to get the fraction we want.

REM To go back divide this same number by a number to

REM get a fraction. This tells us what fraction we want to go back to now

REM add or subtract to get back to this fraction. The deal here is that

REM each fraction has to have only one specific fraction to code to or

REM decode to. Its only these specific fractions that tell us what we

REM need to do.

REM RESULTS OF TESTING....

REM This does work but for each number you use for division you can

```
REM only use half of the results. So if you divide by a specific number

REM and do not get the fraction you want you have to divide by another

REM number until you do. But, you have to make sure you use the same

REM number for decoding. I would suggest starting with a specific
sequence

REM of numbers for division and make sure that when you divide you
get the

REM correct number for decoding each and every time.

REM next create the numbers you can use to test all this stuff.

GOTO prog3:

STOP

prog1:

d# = 1

CLS

DO

PRINT d#; INT(((d# / 4) - INT(d# / 4)) * 10); INT(((d# / 3) - INT(d# / 3))
* 10)

INPUT a$

IF a$ = "s" THEN STOP

d# = d# + 1

LOOP

prog2:

d# = 2

d0# = 1

CLS
```

```
DO

PRINT d0#; INT(((d0# / 4) - INT(d0# / 4)) * 10); d#; INT(((d# / 8) -
INT(d# / 8)) * 10)

INPUT a$

IF a$ = "s" THEN STOP

d# = d# + 2

d0# = d0# + 1

LOOP

prog3:

d1# = 1

d2# = 2

CLS

FOR count = 1 TO 8

PRINT INT(((d1# / 8) - INT(d1# / 8)) * 10); INT(((d2# / 8) - INT(d2# /
8)) * 10)

a$ = INKEY$

REM INPUT a$

IF a$ = "s" THEN STOP

d1# = d1# + 1

d2# = d2# + 2

NEXT count

STOP
```

```
"C:\Users\Reactor1967\vcns\work\080103A.BAS"

DIM can1(1000)

DIM can2(1000)

c1 = 1

c2 = 1

v# = 1

r# = 1

CLS

RANDOMIZE TIMER

DO

z = INT(RND * 2) + 0

IF z = 0 THEN v# = v# + (v# - r# + 1)

IF z = 1 THEN v# = v# + (v# - r# + 2)

PRINT , , z; v#; r#; v# - r#; d#; d# / (v# - r#)

IF z = 0 THEN can1(c1) = d# / (v# - r#)

IF z = 1 THEN can2(c2) = d# / (v# - r#)

IF z = 0 THEN c1 = c1 + 1

IF z = 1 THEN c2 = c2 + 1

IF c1 = 1001 THEN EXIT DO

IF c2 = 1001 THEN EXIT DO

a$ = INKEY$

REM INPUT a$

IF a$ = "s" THEN STOP

d# = INT((v# - r#) * .5)
```

```
r# = v# - d#
IF (r# / 2) - INT(r# / 2) = 0 THEN r# = r# - 1
LOOP
FOR count = 1 TO 1000
PRINT , , can1(count); can2(count)
INPUT a$
IF a$ = "s" THEN STOP
NEXT count
```

```
"C:\Users\Reactor1967\vcns\work\073003B.BAS"
d# = 1
CLS
x# = 1
count = 0
DO
test = ((d# - x#) / 2) - INT((d# - x#) / 2) = 0
r# = x#
IF test = 0 THEN r# = r# + 1
PRINT , , d#; r#; d# - r#
a$ = INKEY$
REM INPUT a$
IF a$ = "s" THEN STOP
```

```
count = count + 1
IF count = 2 THEN x# = x# + 1
IF count = 2 THEN count = 0
d# = d# + 1
LOOP
```

"C:\Users\Reactor1967\vcns\work\073003A.BAS"

```
d1# = 1
d2# = 2
CLS
x# = 1
x2# = 2
count = 0
DO
PRINT , d1#; x#; d1# - x#; " "; d2#; x2#; d2# - x2#
a$ = INKEY$
REM INPUT a$
IF a$ = "s" THEN STOP
d1# = d1# + 2
d2# = d2# + 2
count = count + 1
IF count = 2 THEN x2# = x2# + 2
```

```
IF count = 2 THEN x# = x# + 2
IF count = 2 THEN count = 0
LOOP
```

```
"C:\Users\Reactor1967\vcns\work\072903A.BAS"
v# = 1
r# = 1
rt# = .5
CLS
DO
z = INT(RND * 2) + 0
jack# = v# - r#
IF z = 0 THEN v# = v# + (v# - r# + 1)
IF z = 1 THEN v# = v# + (v# - r# + 2)
PRINT , , z; v#; r#; v# - r#; jack#
a$ = INKEY$
REM INPUT a$
IF a$ = "s" THEN STOP
d# = INT((v# - r#) * rt#)
r# = v# - d#
test = (r# / 2) - INT(r# / 2) = .5
IF test = 0 THEN d# = d# + 1
```

```
r# = v# - d#
LOOP
```

"C:\Users\Reactor1967\vcns\work\072703A.BAS"

```
REM maybe you can divide this into two ways. the distances that you can
REM put data in and the other distances you can not put data in.
REM in the distances you can not put data in use the data stream to
control
REM the code for those distances then read d# at each code to tell if you
can
REM use it divided by a percent to get your previous v# - r# or use the
REM data stream to tell what to do with r#. Its something to think about.
CLS
d# = 1
CLS
DO
x# = INT(d# * .51)
y# = d# - x#
test1 = (y# / 2) - INT(y# / 2) = 0
test2 = INT((x# / .51) + .5) = d#
test3 = (test1 = -1) AND (test2 = -1)
IF test3 = -1 THEN PRINT , d#; INT(d# * .51)
```

```
a$ = INKEY$

REM IF test3 = -1 THEN INPUT a$

IF a$ = "s" THEN STOP


d# = d# + 1

LOOP
```

```
"C:\Users\Reactor1967\vcns\work\072603C.BAS"

REM decode,divide d# / .5. if the previous v# = N of 0 then add 1 to d#

REM right now you have to run the chart in random mode and look for
the

REM value of v# when decoding.

redo:

v# = 5

r# = 1

CLS

RANDOMIZE TIMER

DO

z = INT(RND * 2) + 0

d# = INT((v# - r#) * .5)

r# = v# - d#

x# = v# - r#
```

```
IF z = 0 THEN v# = v# + (v# - r# + 1)

IF z = 1 THEN v# = v# + (v# - r# + 2)

PRINT , , z; v#; r#; v# - r#; x#

IF v# = 14 THEN STOP: REM = 1

IF v# > 14 THEN GOTO redo:

a$ = INKEY$

REM INPUT a$

IF a$ = "s" THEN STOP

test = ((v# - r#) / 2) - INT((v# - r#) / 2) = .5

IF test = -1 THEN r# = r# + 1

LOOP
```

```
"C:\Users\Reactor1967\vcns\work\072603B.BAS"

v# = 5

r# = 1

CLS

RANDOMIZE TIMER

DO

d# = INT((v# - r#) * .5)

r# = r# + ((v# - r#) - d#)

z = INT(RND * 2) + 0

x# = v# - r#
```

```
IF z = 0 THEN v# = v# + (v# - r# + 3)

IF z = 1 THEN v# = v# + (v# - r# + 4)

PRINT , , z; v#; r#; v# - r#; x#; r# - store#

store# = r#

INPUT a$

IF a$ = "s" THEN STOP

LOOP
```

"C:\Users\Reactor1967\vcns\work\072603A.BAS"

```
REM These are the distances that can be used when coding. the only

REM catch is v# = v# + (v# - r#+ (3 or 4)) instead of 1 or 2

REM also you have to start off at the right distance but r# is always

REM odd and d# / .5 = previous v# - r# + or - 1

y# = 1

d# = 1

CLS

DO

test = (d# / 2) - INT(d# / 2)

IF test = .5 THEN z = 0

IF test = 0 THEN z = 1

d2# = INT(d# * .5)

test = (d2# / 2) - INT(d2# / 2)
```

```
IF test = .5 THEN z2 = 0

IF test = 0 THEN z2 = 1

IF (z = z2) AND (INT(d2# / .5) = d#) THEN PRINT , , d#; d2#; d# - d2#

a$ = INKEY$

IF z = z2 AND (INT(d2# / .5) = d#) THEN INPUT a$

IF a$ = "s" THEN STOP

IF z = z2 THEN y# = y# + 1

d# = d# + 1

LOOP UNTIL y# = 1001
```

```
"C:\Users\Reactor1967\vcns\work\072503A.BAS"

v# = 1

r# = 1

CLS

RANDOMIZE TIMER

DO

z = INT(RND * 2) + 0

x# = v# - r#

IF z = 0 THEN v# = v# + (v# - r# + 1)

IF z = 1 THEN v# = v# + (v# - r# + 2)

PRINT z; v#; r#; v# - r#; x#

a$ = INKEY$
```

```
REM INPUT a$

IF a$ = "s" THEN STOP

d# = v# - r#

IF (d# / 2) - INT(d# / 2) = .5 THEN d# = d# - 1

d# = d# / 2

test = ((v# - d#) / 2) - INT((v# - d#) / 2) = .5

IF test = 0 THEN d# = d# + 1

r# = v# - d#

LOOP
```

"C:\Users\Reactor1967\vcns\work\072403A.BAS"

```
REM when you code your data before finding your next v# - r# examine your

REM current value of N. This will tell what ratio you use to find d# = (v# - r#)

REM for your next code. So that, when you decode you look at the value of N for

REM v# and you know what ratio to use to divide into d# to get your previous

REM v# - r#. Anyway you will code using two ratio's and decode using two

REM ratios. Always look at the N value of v# when decoding to see what

REM ratio you use to divide into d# to find your previous v# - r#.

d# = 1
```

```
CLS

DO

a# = INT(.5 * d#)

b# = (d# - a#)

PRINT d#; a#; b#; a# / d#; b# / d#

a$ = INKEY$

INPUT a$

IF a$ = "s" THEN STOP

d# = d# + 1

LOOP
```

```
"C:\Users\Reactor1967\vcns\work\072203D.BAS"

REM Goal of this program is to keep v# - r# even at all times so that

REM if z = 0 then N = 2 and if z = 1 then N = 4. Im searching for a way

REM to map v# - r#;or d# to the value of N of a vector.

REM HOW TO DECODE THIS!!!!!

REM d# is throwing this off because of switching back from even to add.

REM Well d# will alway equal .49 * (v# - r#) take off the fraction

REM speed of v# = (v# - r#) - ((.49 * (v# - r#))

REM v1# - r1# = ((.49 * (v2# - r1#)) / .5

CLS

RANDOMIZE TIMER
```

```
ave# = 0

count = 0

DO

z = INT(RND * 2) + 0

d# = v# - r#

IF z = 0 THEN v# = v# + (v# - r# + 2)

IF z = 1 THEN v# = v# + (v# - r# + 4)

PRINT , z; v#; r#; v# - r#; d#

store# = r#

REM IF v# - r# = 96 THEN EXIT DO

a$ = INKEY$

INPUT a$

IF a$ = "s" THEN STOP

IF a$ = "d" THEN EXIT DO

r# = r# + ((v# - r#) / 2)

LOOP

PRINT ave# / count
```

"C:\Users\Reactor1967\vcns\work\072203C.BAS"

```
REM first program decoding according to 7-20-03 method. Cool.

RANDOMIZE TIMER

DIM can(15000)
```

```
count = 1

v# = 1

r# = 1

CLS

DO

REM z = INT(RND * 2) + 0

REM IF z = 0 THEN v# = v# + (v# - r# + 0)

REM IF z = 1 THEN v# = v# + (v# - r# + 2)

v# = v# + (v# - r# + 1)

test = ((d# / .5) - 1) = store#

PRINT , , v#; r#; v# - r#; d#; test; count

can(count) = v#: count = count + 1

store# = (v# - r#)

a$ = INKEY$

REM INPUT a$

IF a$ = "s" THEN STOP

IF a$ = "d" THEN EXIT DO

IF count = 15000 THEN EXIT DO

d# = (v# - r#)

IF (d# / 2) - INT(d# / 2) = .5 THEN flag = 1 ELSE flag = 0

IF (d# / 2) - INT(d# / 2) = .5 THEN d# = d# - 1

d# = d# / 2

IF flag = 1 THEN d# = d# + 1

r# = v# - d#
```

```
LOOP
INPUT z$
IF z$ = "s" THEN STOP
DO
d# = v# - r#
IF (d# / 2) - INT(d# / 2) = .5 THEN d# = d# - 1
d# = d# / 2
d# = d# / .5
d# = d# - 1
PRINT , , v#; r#; v# - r#; d#; count
test = v# = can(count): count = count - 1
IF test = -1 THEN PRINT , , , "DECODE ERROR!)"
a$ = INKEY$
REM INPUT a$
IF a$ = "s" THEN STOP
dist# = v# - r#
IF (dist# / 2) - INT(dist# / 2) = .5 THEN dist# = dist# - 1
dist# = dist# / 2
dist# = dist# + 1
v# = v# - dist#
r# = v# - d#
IF v# = 1 THEN EXIT DO
LOOP
PRINT , , v#; r#; v# - r#; d#
```

```
"C:\Users\Reactor1967\vcns\work\072203B.BAS"

REM first program decoding according to 7-20-03 method. Cool.

RANDOMIZE TIMER

DIM can(15000)

count = 1

v# = 1

r# = 1

CLS

DO

REM z = INT(RND * 2) + 0

REM IF z = 0 THEN v# = v# + (v# - r# + 0)

REM IF z = 1 THEN v# = v# + (v# - r# + 2)

v# = v# + (v# - r# + 1)

test = ((d# / .5) - 1) = store#

PRINT , , v#; r#; v# - r#; d#; test; count

can(count) = v#: count = count + 1

store# = (v# - r#)

a$ = INKEY$

REM INPUT a$

IF a$ = "s" THEN STOP

IF a$ = "d" THEN EXIT DO
```

```
IF count = 15000 THEN EXIT DO
d# = (v# - r#)
IF (d# / 2) - INT(d# / 2) = .5 THEN flag = 1 ELSE flag = 0
IF (d# / 2) - INT(d# / 2) = .5 THEN d# = d# - 1
d# = d# / 2
IF flag = 1 THEN d# = d# + 1
r# = v# - d#
LOOP
INPUT z$
IF z$ = "s" THEN STOP
DO
d# = v# - r#
IF (d# / 2) - INT(d# / 2) = .5 THEN d# = d# - 1
d# = d# / 2
d# = d# / .5
d# = d# - 1
PRINT , , v#; r#; v# - r#; d#; count
test = v# = can(count): count = count - 1
IF test = -1 THEN PRINT , , , "DECODE ERROR!)"
a$ = INKEY$
REM INPUT a$
IF a$ = "s" THEN STOP
dist# = v# - r#
IF (dist# / 2) - INT(dist# / 2) = .5 THEN dist# = dist# - 1
```

```
dist# = dist# / 2
dist# = dist# + 1
v# = v# - dist#
r# = v# - d#
IF v# = 1 THEN EXIT DO
LOOP
PRINT , , v#; r#; v# - r#; d#
```

```
"C:\Users\Reactor1967\vcns\work\072203A.BAS"
REM first program decoding according to 7-20-03 method. Cool.
RANDOMIZE TIMER
v# = 1
r# = 1
CLS
DO
REM z = INT(RND * 2) + 0
REM IF z = 0 THEN v# = v# + (v# - r# + 0)
REM IF z = 1 THEN v# = v# + (v# - r# + 2)
v# = v# + (v# - r# + 2)
PRINT , , v#; r#; v# - r#; d#
a$ = INKEY$
REM INPUT a$
```

```
IF a$ = "s" THEN STOP

IF a$ = "d" THEN EXIT DO

d# = (v# - r#) * .5

r# = v# - d#

LOOP

DO

PRINT , , v#; r#; v# - r#; d#

IF v# = 1 THEN EXIT DO

a$ = INKEY$

REM INPUT a$

IF a$ = "s" THEN STOP

d# = (v# - r#)

IF (d# / 2) - INT(d# / 2) THEN d# = d# - 1

d# = d# / 2

d# = d# + 1

v# = v# - d#

d# = d# - 2

d# = (d# / .5)

r# = v# - d#

LOOP
```

"C:\Users\Reactor1967\vcns\work\072203.BAS"

```
CLS

RANDOMIZE TIMER

v# = 1

r# = 1

DO

z = INT(RND * 2) + 0

gh# = (v# - r#)

IF z = 0 THEN v# = v# + (v# - r# + 1)

IF z = 1 THEN v# = v# + (v# - r# + 2)

PRINT , z; v#; r#; v# - r#; gh#; r# - st#

st# = r#

a$ = INKEY$

REM INPUT a$

IF a$ = "s" THEN STOP

d# = INT((v# - r#) / 2)

s# = (v# - r#) - d#

IF (s# / 2) - INT(s# / 2) = .5 THEN flag = 1 ELSE flag = 0

IF flag = 1 THEN s# = s# - 1

IF flag = 1 THEN d# = d# + 1

r# = r# + s#

LOOP
```

```
"C:\Users\Reactor1967\vcns\work\072103A.BAS"

DECLARE SUB chart (v#, r#, lb!)

v# = 5

r# = 1

CLS

RANDOMIZE TIMER

DO

lb = INT(RND * 2) + 0

IF lb = 0 THEN v# = v# + (v# - r# + 1)

IF lb = 1 THEN v# = v# + (v# - r# + 2)

PRINT , , lb; v#; r#; v# - r#

a$ = INKEY$

REM INPUT a$

IF a$ = "s" THEN STOP

CALL chart(v#, r#, lb)

LOOP


SUB chart (v#, r#, lb)

control = 1

DO

test = INT(control / 2)

IF (test / 2) - INT(test / 2) = .5 THEN test = test - 1

d# = control

IF (d# / 2) - INT(d# / 2) = .5 THEN d# = d# - 1
```

```
d# = d# / 2

d# = d# + 1

z = (control / 2) - INT(control / 2)

IF z = .5 THEN z = 1

IF z = 0 THEN z1 = 1

IF z = .5 THEN z1 = 0

IF z1 = 0 THEN d# = d# - 1

IF z1 = 1 THEN d# = d# - 2

REM PRINT "dist= "; d#; "U'r v# - r#= "; control; "U'r spd of r#= "; test;
"= "; "Prev. v# - r#= "; d# + test

REM PRINT , , d#; control; test; " = "; d# + test

IF (control / 2) - INT(control / 2) = .5 THEN sheila = 0

IF (control / 2) - INT(control / 2) = 0 THEN sheila = 1

Meredith = ((d# + test) = (v# - r#)) AND (sheila = lb)

IF Meredith = -1 THEN EXIT DO

a$ = INKEY$

REM INPUT a$

IF a$ = "s" THEN STOP

control = control + 1

LOOP

dist# = control

IF (dist# / 2) - INT(dist# / 2) = .5 THEN dist# = dist# - 1

dist# = dist# / 2

dist# = dist# + 1
```

```
IF lb = 0 THEN dist# = dist# - 1

IF lb = 1 THEN dist# = dist# - 2

r# = v# - dist#

END SUB
```

```
"C:\Users\Reactor1967\vcns\work\072003E.BAS"

REM assuming your speed of r# is half your current value of v# - r#

REM you can compute your dist before coding with your current v# - r#

REM add your dist + your speed of r# to = the v# - r# you came from

REM now use this has a chart to code/decode data with. L.B.

REM so the final answer to getting vcns to code is know the ratio

REM of your speed of r# to dist before coding this use that to
mathmatically

REM create a code and decode chart for your v# - r#;speed of r#; and
distance.

control = 1

CLS

DO

test = INT(control / 2)

IF (test / 2) - INT(test / 2) = .5 THEN test = test - 1

d# = control

IF (d# / 2) - INT(d# / 2) = .5 THEN d# = d# - 1

d# = d# / 2
```

```
d# = d# + 1

z = (control / 2) - INT(control / 2)

IF z = .5 THEN z = 1

IF z = 0 THEN z1 = 1

IF z = .5 THEN z1 = 0

IF z1 = 0 THEN d# = d# - 1

IF z1 = 1 THEN d# = d# - 2

REM PRINT "dist= "; d#; "U'r v# - r#= "; control; "U'r spd of r#= "; test;
"= "; "Prev. v# - r#= "; d# + test

PRINT , , d#; control; test; " = "; d# + test

a$ = INKEY$

INPUT a$

IF a$ = "s" THEN STOP

control = control + 1

LOOP UNTIL control > 200

REM Getting v# and R# closer. See if r# increases faster than v# then

REM v# and r# get closer and if v# increases faster than r# v# and r# get

REM farther apart. Use this in the charts.

DO

test = INT(control * .75)

IF (test / 2) - INT(test / 2) = .5 THEN test = test - 1

d# = control

IF (d# / 2) - INT(d# / 2) = .5 THEN d# = d# - 1

d# = d# / 2
```

```
d# = d# + 1

z = (control / 2) - INT(control / 2)

IF z = .5 THEN z = 1

IF z = 0 THEN z1 = 1

IF z = .5 THEN z1 = 0

IF z1 = 0 THEN d# = d# - 1

IF z1 = 1 THEN d# = d# - 2

REM PRINT "dist= "; d#; "U'r v# - r#= "; control; "U'r spd of r#= "; test; "= "; "Prev. v# - r#= "; d# + test

PRINT , , d#; control; test; " = "; d# + test

a$ = INKEY$

INPUT a$

IF a$ = "s" THEN STOP

control = control - 1

LOOP UNTIL control = 1
```

```
"C:\Users\Reactor1967\vcns\work\072003D.BAS"

c# = 1000

RANDOMIZE TIMER

CLS

DO
```

```
store# = c#

test = (c# / 2) - INT(c# / 2)

IF test = 0 THEN z1 = 1

IF test = .5 THEN z1 = 0

z = INT(RND * 2) + 0

IF z = z1 THEN c# = c# + 2

IF z <> z1 THEN c# = c# + 1

dist# = c#

IF (dist# / 2) - INT(dist# / 2) = .5 THEN dist# = dist# - 1

dist# = dist# / 2

dist# = dist# + 1

IF z = 0 THEN dist# = dist# - 1

IF z = 1 THEN dist# = dist# - 2

s# = store# - dist#

PRINT , , z; c#; dist#; s#

a$ = INKEY$

IF a$ = "s" THEN STOP

LOOP
```

"C:\Users\Reactor1967\vcns\work\072003C.BAS"

```
REM TRYING TO FIND MY THING HERE. IF yOU KNOW YOUR SPEED OF R#, YOUR DIST
```

```
REM BEFORE CODING, AND YOUR Z CAN YOU CALCULATE WHEN DECODING WHAT YOUR
REM PREVIOUS DIST AND SPEED OF R IS BY LOOKING AT YOUR VALUES OF Z WHEN
REM DECODING?
vr# = 1000
RANDOMIZE TIMER
CLS
DO
z = INT(RND * 2) + 0
store# = vr#
vr# = vr# + 1
t# = (vr# / 2) - INT(vr# / 2)
test1 = (z = 0) AND (t = .5)
test2 = (z = 1) AND (t = 0)
test3 = (test1 = -1) OR (test2 = -1)
IF test3 = 0 THEN vr# = vr# + 1
s# = vr#
IF (s# / 2) - INT(s# / 2) = .5 THEN s# = s# - 1
s# = s# / 2
IF (s# / 2) - INT(s# / 2) = .5 THEN s# = s# - 1
dist# = store# - s#
IF z = 0 THEN vr# = (dist# * 2) + 1
IF z = 1 THEN vr# = (dist# * 2) + 2
PRINT , , z; vr#; dist#; dist# - store1#; s#; s# - store2#
```

```
store1# = dist#
store2# = s#
a$ = INKEY$
IF a$ = "s" THEN STOP
LOOP
```

"C:\Users\Reactor1967\vcns\work\072003B.BAS"

```
d# = 1000
v# = 10001
r# = v# - d#
CLS
DO
c# = d#
z = INT(RND * 2) + 0
DO
c# = c# + 1
test = (c# / 2) - INT(c# / 2)
test2 = (z = 1) AND (test = 0)
test3 = (z = 0) AND (test = .5)
x# = c#
IF (x# / 2) - INT(x# / 2) = .5 THEN x# = x# - 1
x# = x# / 2
```

```
x# = x# + 1

IF z = 0 THEN x# = x# - 1

IF z = 1 THEN x# = x# - 2

test4 = ((v# - x#) / 2) - INT((v# - x#) / 2) = .5

test5 = ((test2 = -1) OR (test3 = -1)) AND (test4 = -1)

IF test5 = -1 THEN EXIT DO

a$ = INKEY$

IF a$ = "s" THEN STOP

LOOP

r# = v# - x#

v# = r# + c#

PRINT , , z; v#; r#; v# - r#; r# - store1#

store1# = r#

INPUT a$

IF a$ = "s" THEN STOP

d# = c#

LOOP
```

```
"C:\Users\Reactor1967\vcns\work\072003A.BAS"

DIM can(3000)

DIM cam(3000)

CLS
```

```
RANDOMIZE TIMER
count2 = 1
redo:
v# = 1
r# = 1
DO
z = INT(RND * 2) + 0
IF z = 0 THEN v# = v# + (v# - r# + 1)
IF z = 1 THEN v# = v# + (v# - r# + 2)
PRINT , , z; v#; r#; v# - r#; r# - store1#; count2
dist# = v# - r#
IF dist# > 1000 THEN EXIT DO
FOR count = 1 TO 3000
test = (can(count) = dist#) AND (cam(count) = r# - store1#)
IF test = -1 THEN EXIT FOR
NEXT count
IF test = 0 THEN cam(count2) = r# - store1#
IF test = 0 THEN can(count2) = dist#
IF test = 0 THEN count2 = count2 + 1
store1# = r#
REM store# = v# - r#
a$ = INKEY$
REM INPUT a$
IF a$ = "s" THEN STOP
```

```
IF z = 0 THEN dist# = dist# - 1
IF z = 1 THEN dist# = dist# - 2
test = (dist# / 2) - INT(dist# / 2) = .5
IF test = -1 THEN dist# = dist# - 1
dist# = dist# / 2
test = (dist# / 2) - INT(dist# / 2) = .5
IF test = -1 THEN dist# = dist# - 1
r# = r# + dist#
LOOP
a$ = INKEY$
IF a$ = "s" THEN STOP
GOTO redo:
```

```
"C:\Users\Reactor1967\vcns\work\071903A.BAS"
REM test should be v# (decoded with r2#) => r2# & v# (decoded with
r1#) < r2# and => r1#
REM if test = -1 then r2# = r1#, if test = 0 then r2# = r2#
REM question, can this be used to coded and decode with?
REM new equation v# = v# + ((v# - r#) * (base - 1)) + (N - 1
REM Kicker to this new equation you can,t repeat 0's but here is what
you
REM can do code 10 as a binary 0 and 11 as a binary 1. With the way this
REM this equation sets up R# is easier to keep up with and there is no
```

```
REM runaway as when your coding N instead of N - 1. Now write a
program

REM that goes with it.

v# = 1

r# = 1

RANDOMIZE TIMER

CLS

x# = 101

DO

z = INT(RND * 2) + 0

IF z = 0 THEN v# = v# + (v# - r# + 0)

IF z = 0 THEN PRINT , z; v#; r#;

test = v# >= x#

IF test = -1 THEN r# = r# + 100

IF test = -1 THEN x# = x# + 100

IF z = 0 THEN v# = v# + (v# - r# + 1)

IF z = 0 THEN PRINT " "; v#; r#

test = v# >= x#

IF test = -1 THEN r# = r# + 100

IF test = -1 THEN x# = x# + 100

IF z = 1 THEN v# = v# + (v# - r# + 1)

IF z = 1 THEN PRINT , z; v#; r#;

test = v# >= x#

IF test = -1 THEN r# = r# + 100
```

```
IF test = -1 THEN x# = x# + 100

IF z = 1 THEN v# = v# + (v# - r# + 1)

IF z = 1 THEN PRINT " "; v#; r#

test = v# >= x#

IF test = -1 THEN r# = r# + 100

IF test = -1 THEN x# = x# + 100

a$ = INKEY$

REM INPUT a$

IF a$ = "s" THEN STOP

LOOP
```

```
"C:\Users\Reactor1967\vcns\work\071803D.BAS"

REM Demonstration of the Vector Coordinate Numerical System.

v# = 1

r# = 1

v2# = 1

r2# = 1

hgj# = 1

CLS

by# = 0

RANDOMIZE TIMER

z = 0
```

```
TIMER ON
a1$ = TIME$
DO
z = INT(RND * 2) + 0
IF r# < 0 THEN r# = 1
IF r# > v# THEN r# = r# - 100
IF z = 0 THEN v# = v# + (v# - r# + 1)
IF z = 1 THEN v# = v# + (v# - r# + 2)
by# = by# + 1
REM PRINT , , by#; z; v#; r#; v# - r#; v# - store1#
PRINT , , , z; r#
store1# = v#
a$ = INKEY$
INPUT a$
IF a$ = "s" THEN STOP
IF a$ = "d" THEN EXIT DO
sr1# = r#
t# = (v# / 100) - INT(v# / 100)
r# = (v# / 100) - t#
r# = (r# * 100) + 1
sr2# = r#
REM IF sr1# = sr2# THEN z = 0
REM IF sr1# <> sr2# THEN z = 1
LOOP
```

```
b$ = TIME$

PRINT "Encode Complete! Number of bits encoded is "; by#; " Hit Enter to begin decode."

PRINT "starting coding at "; a1$

PRINT "stoped coding at "; b$

INPUT z$

IF z$ = "s" THEN SYSTEM

sby# = by#

REM ------------------------------------------------

REM STOP

c$ = TIME$

DO

test = (v# / 2) - INT(v# / 2)

IF test = 0 THEN z = 0

IF test = .5 THEN z = 1

PRINT , , by#; z; v#; r#; v# - r#

by# = by# - 1

a$ = INKEY$

REM INPUT a$

IF a$ = "s" THEN STOP

dist# = v# - r#

IF (dist# / 2) - INT(dist# / 2) = .5 THEN dist# = dist# - 1

dist# = dist# / 2

dist# = dist# + 1
```

```
v# = v# - dist#

IF z = 1 THEN r# = r# - 100

IF r# <= 0 THEN r# = 1

IF v# <= 1 THEN d$ = TIME$

IF v# <= 1 THEN PRINT , , by#; z; v#; r#; v# - r#

IF v# <= 1 THEN PRINT "Decode Complete!"

IF v# <= 1 THEN PRINT "started coding at "; a1$; " stoped coding at ";
b$

IF v# <= 1 THEN PRINT "started decoding at "; c$; " stoped decoding at
"; d$

IF v# <= 1 THEN PRINT sby#; " Bits decoded."

IF v# <= 1 THEN INPUT z$

IF v# <= 1 THEN SYSTEM

LOOP
```

```
"C:\Users\Reactor1967\vcns\work\071803C.BAS"

REM maybe a better approach would be to write a program that

REM repeats over and over and charts the high and low values

REM of v# for each value of r# and see where this leads. Have

REM the program run at different low hi values of v# until

REM a specific region of r's are chart then move up to rem different

REM values of v#.

RANDOMIZE TIMER
```

```
DIM can1(1000)
DIM can2(1000)
ON ERROR GOTO ouy:
z = 0
db = 1
redo:
CLS
v# = 1
r# = 1
DO
z = INT(RND * 2) + 0
IF r# < 0 THEN r# = 1
IF r# > v# THEN r# = r# - 100
IF z = 0 THEN v# = v# + (v# - r# + 1)
IF z = 1 THEN v# = v# + (v# - r# + 2)
PRINT , , z; v#; r#; v# - r#; db
a$ = INKEY$
REM INPUT a$
IF a$ = "s" THEN STOP
IF a$ = "d" THEN GOTO ouy:
FOR db2 = 1 TO 1000
test = (v# = can1(db2))
IF test = -1 THEN EXIT FOR
NEXT db2
```

```
IF test = 0 THEN can1(db) = v#

IF test = 0 THEN can2(db) = r#

IF test = 0 THEN db = db + 1

t# = (v# / 100) - INT(v# / 100)

r# = (v# / 100) - t#

r# = (r# * 100) + 1

LOOP UNTIL r# > 100

GOTO redo:

ouy:

FOR db = 1 TO 1000

PRINT , , can1(db); can2(db)

INPUT a$

IF a$ = "s" THEN STOP

NEXT db
```

```
"C:\Users\Reactor1967\vcns\work\071803B.BAS"

REM Demonstration of the Vector Coordinate Numerical System.

REM program failed test results were inconcluvise due to repeative
results

REM yealding different values for speed of r#.

REM It seems some type of weight number will need to be compututed as
the
```

```
REM the program codes and that weighted number will tell the speed of r#.

REM the process will needed to be reveresed when decoded and the weighted

REM number will have to stay within bounderies perhaps going up and down in

REM value itself with the + and - ends telling which way to compute the

REM weight.

REM -----------------------------------------------------------

REM maybe a better approach would be to write a program that

REM repeats over and over and charts the high and low values

REM of v# for each value of r# and see where this leads. Have

REM the program run at different low hi values of v# until

REM a specific region of r's are chart then move up to rem different

REM values of v#.

DIM can1(500)

DIM can2(500)

DIM can3(500)

DIM can4(500)

DIM can5(500)

db2 = 1

v# = 274

r# = 101

v2# = 1

r2# = 1
```

```
hgj# = 1

CLS

by# = 0

RANDOMIZE TIMER

z = 0

TIMER ON

a1$ = TIME$

DO

zstore = z

z = INT(RND * 2) + 0

IF r# < 0 THEN r# = 1

IF r# > v# THEN r# = r# - 100

ghwells# = (v# - r#)

IF z = 0 THEN v# = v# + (v# - r# + 1)

IF z = 1 THEN v# = v# + (v# - r# + 2)

by# = by# + 1

PRINT , , by#; z; v#; r#; v# - r#; zstore; ghwells#; r# - store2#; db2

FOR db = 1 TO 500

test = (z = can1(db)) AND (zstore = can2(db)) AND (ghwells# =
can3(db)) AND (r# - store2# = can4(db)) AND ((v# - r#) = can5(db))

IF test = -1 THEN EXIT FOR

NEXT db

IF test = 0 THEN can1(db2) = z

IF test = 0 THEN can2(db2) = zstore
```

```
IF test = 0 THEN can3(db2) = ghwells#

IF test = 0 THEN can4(db2) = r# - store2#

IF test = 0 THEN can5(db2) = v# - r#

IF test = 0 THEN db2 = db2 + 1

store2# = r#

a$ = INKEY$

INPUT a$

IF a$ = "s" THEN STOP

IF a$ = "d" THEN EXIT DO

t# = (v# / 100) - INT(v# / 100)

r# = (v# / 100) - t#

r# = (r# * 100) + 1

LOOP

FOR db = 1 TO 407

PRINT , , can1(db); can2(db); can3(db); can4(db); can5(db)

INPUT a$

IF a$ = "s" THEN STOP

NEXT db
```

```
"C:\Users\Reactor1967\vcns\work\071803A.BAS"

REM Demonstration of the Vector Coordinate Numerical System.

REM studying bits coding at to bits coded from, d# before coding, and
```

```
REM speed of r# after coding. Looking for pattern.

v# = 1

r# = 1

v2# = 1

r2# = 1

hgj# = 1

CLS

by# = 0

RANDOMIZE TIMER

z = 0

TIMER ON

a1$ = TIME$

DO

REM z = INT(RND * 2) + 0

IF r# < 0 THEN r# = 1

IF r# > v# THEN r# = r# - 100

ghwells# = v# - r#

IF z = 0 THEN v# = v# + (v# - r# + 1)

IF z = 1 THEN v# = v# + (v# - r# + 2)

by# = by# + 1

PRINT , by#; z; v#; r#; v# - r#; v# - store1#; ghwells#; r# - store2#

IF ghwells# = 67 THEN STOP

store1# = v#

store2# = r#
```

```
a$ = INKEY$

REM INPUT a$

IF a$ = "s" THEN STOP

IF a$ = "d" THEN EXIT DO

sr1# = r#

t# = (v# / 100) - INT(v# / 100)

r# = (v# / 100) - t#

r# = (r# * 100) + 1

sr2# = r#

IF sr1# = sr2# THEN z = 0

IF sr1# <> sr2# THEN z = 1

LOOP

b$ = TIME$

PRINT "Encode Complete! Number of bits encoded is "; by#; " Hit Enter
to begin decode."

PRINT "starting coding at "; a1$

PRINT "stoped coding at "; b$

INPUT z$

IF z$ = "s" THEN SYSTEM

sby# = by#

REM -----------------------------------------------

REM STOP

c$ = TIME$

DO
```

```
test = (v# / 2) - INT(v# / 2)

IF test = 0 THEN z = 0

IF test = .5 THEN z = 1

PRINT , , by#; z; v#; r#; v# - r#

by# = by# - 1

a$ = INKEY$

REM INPUT a$

IF a$ = "s" THEN STOP

dist# = v# - r#

IF (dist# / 2) - INT(dist# / 2) = .5 THEN dist# = dist# - 1

dist# = dist# / 2

dist# = dist# + 1

v# = v# - dist#

IF z = 1 THEN r# = r# - 100

IF r# <= 0 THEN r# = 1

IF v# <= 1 THEN d$ = TIME$

IF v# <= 1 THEN PRINT , , by#; z; v#; r#; v# - r#

IF v# <= 1 THEN PRINT "Decode Complete!"

IF v# <= 1 THEN PRINT "started coding at "; a1$; " stoped coding at ";
b$

IF v# <= 1 THEN PRINT "started decoding at "; c$; " stoped decoding at
"; d$

IF v# <= 1 THEN PRINT sby#; " Bits decoded."

IF v# <= 1 THEN INPUT z$

IF v# <= 1 THEN SYSTEM
```

```
LOOP




"C:\Users\Reactor1967\vcns\work\071203D.BAS"
DIM can(1000)
c1# = 1
c2# = 1
d# = 0
v# = 1
r# = 1
CLS
DO
y# = (v# / 2) - INT(v# / 2)
IF y# = 0 THEN z = 0
IF y# = .5 THEN z = 1
PRINT , , z; v# - r#; (v# + (v# - r# + 1)) - r#; (v# + (v# - r# + 2)) - r#; c2#
ash# = v# - r#
INPUT a$
IF a$ = "s" THEN STOP
test = (v# / 2) - INT(v# / 2)
IF test = 0 THEN can(c1#) = (ash# + 2)
IF test = .5 THEN can(c1#) = (ash# - r# + 1)
c1# = c1# + 1
```

```
IF test = 0 THEN can(c1#) = (ash# - r#) + 3

IF test = .5 THEN can(c1#) = (ash# - r#) + 4

c1# = c1# + 1

v# = can(c2#)

c2# = c2# + 1

d# = d# + 1

r# = v# - d#

LOOP
```

"C:\Users\Reactor1967\vcns\work\071203C.BAS"

```
REM this program uses 071203b chart

REM Now a new ideal. Just as distance from v# - r# can tell us N maybe

REM setting up a system so that v# - r# after coding tells us our speed

REM of r#. Hummm that sounds interesting. If distance tells us our speed

REM of r# then I can just use my math equations to find N.

REM so speed of r# = (v1# - r1#) - (v1# - r2#) so I need to

REM setup specific distances for (v# - r#) after coding for specific

REM next N values to be coded. Then put all this in a chart so that when

REM I decode I can compare v# - r# before decoding to know my
previous

v# = 1000

r# = 999
```

```
z = 1

d1# = 1: REM binary 0

d2# = 2: REM binary 1

CLS

RANDOMIZE TIMER

DO

IF z = 0 THEN v# = v# + (v# - r#) + 1

IF z = 1 THEN v# = v# + (v# - r#) + 2

PRINT , , z; v#; r#; v# - r#; (v# - r#) - storage#

storage# = (v# - r#)

a$ = INKEY$

INPUT a$

IF a$ = "s" THEN STOP

d1# = d1# + 2

d2# = d2# + 2

IF z = 0 THEN r# = v# - d1#

IF z = 1 THEN r# = v# - d2#

z = INT(RND * 2) + 0

LOOP
```

"C:\Users\Reactor1967\vcns\work\071203B.BAS"

REM !!!!!!!!!!!!!!!!!!!!!!!!!!!!IMPORTANT BREAKTHROUGH!!!!!!!!!!!!!!!!!!!!!

REM I,ve don this program before a long time ago and here I am again back

REM at it but this time I know something different. Fist if you know the

REM distance your going to code before hand you can control where v and r

REM get closer or farther apart. I knew that before. Here is what I learn

REM that is new. A few weeks ago I realized that the distance v is from

REM r can be used as an indicator to tell the n value of v. With this

REM in mind I can use the distance to tell N and work on knowing what

REM Im coding to depending if im going from a 0 to a 0 or a 0 to a 1

REM or a 1 to a 0 or a 1 to a 0. N will only have two choices to code

REM to but distances will have base^2 squared choices.

REM now here is how all this helps me. When I decode a vector or v#

REM I look at v# - r# before I decode and that tells me the N value

REM of v#. I then decode V# and look at the v# - r# again with the

REM same R# and that tells me again what N is for the value v# is

REM currently at. So I know N before hand and I know N after the fact.

REM Now I can take those two N values and know what my previous v# - r#

REM should be. This program is as close as I can get right now to

REM a prototype of that type of decode system.

```
REM use this chart for the next program as a tool for coding and decoding.

REM -----------------------------------------------------------------

REM rules if v - r = odd then n = 1

REM rules if v - r = even then n = 2

REM rules binary 0 coded to binary 0 (v - r) = (v - r) + 4

REM rules binary 1 coded to binary 1 (v - r) = (v - r) + 4

REM rules binary 0 coded to binary 1 (v - r) = (v - r) + 5

REM rules binary 1 coded to binary 0 (v - r) = (v - r) + 5

REM rules previous (v - r) = ((even v - r) / 2) - 1

REM rules previous (v - r) = ((odd v - r) - 1) / 2

REM -----------------------------------------------------------------


t1 = 0

t2 = 1

d1# = 1

d2# = 2

v# = 1000

CLS

DO

r1# = v# - d1#

r2# = v# - d2#

PRINT , , t1; v# - r1#; (v# + (v# - r1# + 1)) - r1#; (v# + (v# - r1# + 2)) - r1#; "|"; d1#
```

```
PRINT , , t2; v# - r2#; (v# + (v# - r2# + 1)) - r2#; (v# + (v# - r2# + 2)) - r2#; "|"; d2#

d1# = d1# + 2

d2# = d2# + 2

a$ = INKEY$

INPUT a$

IF a$ = "s" THEN STOP

LOOP
```

```
"C:\Users\Reactor1967\vcns\work\071203A.BAS"

REM I,ve don this program before a long time ago and here I am again back

REM at it but this time I know something different. Fist if you know the

REM distance your going to code before hand you can control where v and r

REM get closer or farther apart. I knew that before. Here is what I learn

REM that is new. A few weeks ago I realized that the distance v is from

REM r can be used as an indicator to tell the n value of v. With this

REM in mind I can use the distance to tell N and work on knowing what

REM Im coding to depending if im going from a 0 to a 0 or a 0 to a 1

REM or a 1 to a 0 or a 1 to a 0. N will only have two choices to code

REM to but distances will have base^2 squared  choices.
```

```
REM now here is how all this helps me. When I decode a vector or v#
REM I look at v# - r# before I decode and that tells me the N value
REM of v#. I then decode V# and look at the v# - r# again with the
REM same R# and that tells me again what N is for the value v# is
REM currently at. So I know N before hand and I know N after the fact.
REM Now I can take those two N values and know what my previous v#
- r#
REM should be. This program is as close as I can get right now to
REM a prototype of that type of decode system.


v# = 10
r# = 9
v# = v# + (v# - r# + 2)
d# = v# - r#
z = 0
CLS
RANDOMIZE TIMER
DO
IF v# - r# < 50 THEN flag = 1
IF v# - r# > 90 THEN flag = 0
PRINT , , z; v#; r#; v# - r#; d#; d# - store#
store# = d#
a$ = INKEY$
REM INPUT a$
```

```
IF a$ = "s" THEN STOP

z = INT(RND * 2) + 0

DO

IF flag = 1 THEN d# = d# + 1

IF flag = 0 THEN d# = d# - 1

t# = d#

IF (t# / 2) - INT(t# / 2) = .5 THEN t# = t# - 1

t# = t# / 2

t# = t# + 1

IF z = 0 THEN t# = t# - 1

IF z = 1 THEN t# = t# - 2

r# = v# - t#

test = (r# / 2) - INT(r# / 2) = .5

a$ = INKEY$

IF a$ = "s" THEN STOP

LOOP UNTIL test = -1

IF z = 0 THEN v# = v# + (v# - r# + 1)

IF z = 1 THEN v# = v# + (v# - r# + 2)

LOOP
```

"C:\Users\Reactor1967\vcns\work\070603A.BAS"

```
REM something to play with. Here an independant variable is used to control

REM the incrementing of r#. It works for a bit then breaks containment.

v# = 1

R# = 1

tag = 0

CLS

RANDOMIZE TIMER

ON ERROR GOTO ed:

DO

z = INT(RND * 2) + 0

IF z = 0 THEN v# = v# + (v# - R# + 1)

IF z = 1 THEN v# = v# + (v# - R# + 2)

PRINT , , v#; R#; v# - R#; tag

IF v# - R# > 200 THEN PRINT "Containment Breach your program is going to crash."

IF v# > 999999999999999# THEN PRINT , , "     BOOM!!!!!!!!!!!!!!!!"

IF R# > 999999999999999# THEN PRINT , , "     BOOM!!!!!!!!!!!!!!!!"

tag = tag * 2

IF z = 0 THEN tag = tag + 1

IF z = 1 THEN tag = tag + 2

IF tag >= 99 THEN R# = R# + 100

IF tag >= 99 THEN tag = tag - 100

IF tag <= 0 THEN tag = 1

IF R# > v# THEN R# = R# - 100
```

```
A$ = INKEY$
IF v# < 999999999999999# THEN INPUT A$
IF A$ = "s" THEN STOP
LOOP
ed:
CLS
PRINT , , "  YOUR PROGRAM BLEW SKI HIGH!!!!"
INPUT A$
SYSTEM
```

```
"C:\Users\Reactor1967\vcns\work\070506A.BAS"
v# = 1
r# = 1
CLS
RANDOMIZE TIMER
z = 0
DO
REM z = INT(RND * 2) + 0
IF z = 0 THEN v# = v# + (v# - r# + 1)
IF z = 1 THEN v# = v# + (v# - r# + 2)
PRINT , , z; v#; r#; v# - r#
store1# = r#
```

```
a$ = INKEY$
REM INPUT a$
IF a$ = "s" THEN STOP
IF a$ = "d" THEN EXIT DO
sr1# = r#
t# = (v# / 100) - INT(v# / 100)
r# = (v# / 100) - t#
r# = (r# * 100) + 1
IF r# <= 0 THEN r# = 1
sr2# = r#
IF sr1# = sr2# THEN z = 0
IF sr1# <> sr2# THEN z = 1
LOOP
DO
test = (v# / 2) - INT(v# / 2)
IF test = 0 THEN z = 0
IF test = .5 THEN z = 1
PRINT , , z; v#; r#; v# - r#
a$ = INKEY$
REM INPUT a$
IF a$ = "s" THEN STOP
dist# = v# - r#
IF (dist# / 2) - INT(dist# / 2) = .5 THEN dist# = dist# - 1
dist# = dist# / 2
```

```
dist# = dist# + 1

v# = v# - dist#

IF z = 1 THEN r# = r# - 100

IF r# <= 0 THEN r# = 1

IF v# <= 1 THEN STOP

LOOP
```

"C:\Users\Reactor1967\vcns\work\070503B.BAS"

```
v# = 1

r# = 1

DIM can1(1000)

count1 = 1

count2 = 1

CLS

DO

store1# = r#

a# = v#

b# = v# + (v# - r# + 1)

c# = v# + (v# - r# + 2)

IF v# >= (r# + 99) THEN r# = r# + 100

can1(count1) = b#

count1 = count1 + 1
```

```
can1(count1) = c#
count1 = count1 + 1
v# = can1(count2)
r# = 1
DO
sr# = r#
r# = r# + 100
IF r# > v# THEN EXIT DO
a$ = INKEY$
IF a$ = "s" THEN STOP
LOOP
r# = sr#
count2 = count2 + 1
PRINT , , a#; store1#; "|"; b#; c#
a$ = INKEY$
REM INPUT a$
IF a$ = "s" THEN STOP
LOOP UNTIL count1 >= 1000
```

"C:\Users\Reactor1967\vcns\work\070503A.BAS"

```
REM this does not work. wrong approach.

v# = 1

r# = 1

CLS

DO

v2# = v# + (v# - r# + 1)

v3# = v# + (v# - r# + 2)

IF (v2# - r#) >= (r# + 99) THEN r2# = r# + 100 ELSE r2# = r#

IF (v3# - r#) >= (r# + 99) THEN r3# = r# + 100 ELSE r3# = r#

PRINT , , v#; r#; "|"; v2#; r2#; "|"; v3#; r3#

v# = v# + 1

IF v# - r# >= (r# + 99) THEN r# = r# + 100

a$ = INKEY$

INPUT a$

IF a$ = "s" THEN STOP

LOOP
```

"C:\Users\Reactor1967\vcns\work\070403A.BAS"

```
v# = 1

r# = 1

DIM can1(1000)

DIM can2(1000)
```

```
DIM can3(1000)

CLS

RANDOMIZE TIMER

count = 1

DO

z = INT(RND * 2) + 0

a# = v# - r#

IF z = 0 THEN v# = v# + (v# - r# + 1)

IF z = 1 THEN v# = v# + (v# - r# + 2)

PRINT , , z; v#; r#; v# - r#; a#; b#; count

a$ = INKEY$

REM INPUT a$

IF a$ = "s" THEN EXIT DO

test = v# >= (r# + 99)

IF test = -1 THEN r# = r# + 100

b# = v# - r#

FOR shuttle = 1 TO 1000

test = (a# = can1(shuttle)) OR (b# = can2(shuttle))

IF test = -1 THEN EXIT FOR

NEXT shuttle

IF test = 0 THEN can1(count) = a#

IF test = 0 THEN can2(count) = b#

IF test = 0 THEN count = count + 1

LOOP
```

```
count = 1
FOR count = 1 TO 100
PRINT , , can1(count); can2(count)
INPUT a$
IF a$ = "s" THEN STOP
NEXT count
```

"C:\Users\Reactor1967\vcns\work\062903B.BAS"

```
DECLARE SUB findvalue (v#, r#, z!)
v# = 3
r# = 1
RANDOMIZE TIMER
CLS
DO
z1 = INT(RND * 2) + 0
test# = (v# - r#)
IF z1 = 0 THEN v# = v# + (v# - r# + 2)
IF z1 = 1 THEN v# = v# + (v# - r# + 4)
CALL findvalue(v#, r#, z)
PRINT , , z; z1; v#; r#; v# - r#; v# - store1#; r# - store2#
store1# = v#
```

```
store2# = r#
a$ = INKEY$
INPUT a$
IF a$ = "s" THEN STOP
r# = r# + test#
LOOP

SUB findvalue (v#, r#, z)
d# = 1
v1# = 9999998
DO
r1# = v1# - d#
a# = v1# + (v1# - r1# + 2)
b# = v1# + (v1# - r1# + 4)
test1 = ((a# - r1#) = (v# - r#))
test2 = ((b# - r1#) = (v# - r#))
IF test1 = -1 THEN z = 1
IF test2 = -1 THEN z = 0
test3 = (test1 = -1) OR (test2 = -1)
IF test3 = -1 THEN EXIT DO
a$ = INKEY$
IF a$ = "s" THEN STOP
d# = d# + 2
LOOP
```

```
END SUB

"C:\Users\Reactor1967\vcns\work\062903A.BAS"

REM goal here is to keep everything even so I can work on speed of r#

REM I did not work on speed of r# yet just starting some test.

v# = 1000

d# = 1

CLS

DO

r# = v# - d#

PRINT , , (v# + (v# - r# + 2)); (v# + (v# - r# + 4)); r#; (v# + (v# - r# + 2)) - r#; (v# + (v# - r# + 4)) - r#

INPUT a$

IF a$ = "s" THEN STOP

d# = d# + 2

LOOP

"C:\Users\Reactor1967\vcns\work\062803D.BAS"

v# = 1
```

```
r# = 1

CLS

RANDOMIZE TIMER

DO

z = INT(RND * 2) + 0

IF test = -1 THEN z = 2

IF r# > v# THEN z = 2

IF z = 0 THEN v# = v# + (v# - r# + 1)

IF z = 1 THEN v# = v# + (v# - r# + 2)

IF z = 2 THEN v# = v# + (v# - r# + 3)

PRINT , , z; v#; r#; v# - r#

IF r# > v# THEN STOP

a$ = INKEY$

REM INPUT a$

IF a$ = "s" THEN STOP

IF (v# - r#) >= 100 THEN r# = r# + 100

LOOP
```

```
"C:\Users\Reactor1967\vcns\work\062803C.BAS"

REM attempting to create some type of chart so then when you decode
you
```

```
REM look at your binary numbers such as a 0 decoding to a 0 or a 0
REM decoding to a 1 or a 1 decoding to a 0 or a 1 decoding to a 1
REM to put on a chart and look to see what your v# - r# is for your
REM decoded number. It might be possible to go up and down the chart
REM in this fashion coding data to a specific v# - r# knowing that
REM when you decode just take know your position on the chart and
REM take your binary data to find your previous v# - r#
REM No method has been developed yet to do this.
REM Important thing learned here is that for in each base for every
REM distance used before coding you have the number of possibilites
REM equal to the base for distances(v# - r#) after coding and it is
REM possible to create a chart expressing this. Now, maybe it might
REM be possible to estabish a patter for coding data using that chart.
REM Next now that you can create a chart for you v - r's to the same
REM with your speed of v's and speed of r's see if you learn anything.
v1# = 1000
v2# = 1001
d# = 1
d2# = 2
CLS
vt# = 1000
RANDOMIZE TIMER
DO
t1# = vt# - d#
```

```
t2# = vt# - d2#

IF (t1# / 2) - INT(t1# / 2) = .5 THEN rt# = t1#

IF (t2# / 2) - INT(t2# / 2) = .5 THEN rt# = t2#

z = INT(RND * 2) + 0

vt# = vt# + (vt# - rt#) + z + 1

v# = v1#

r# = v# - d#

PRINT (v# + (v# - r# + 1)) - r#; (v# + (v# - r# + 2)) - r#;

v# = v2#

r# = v# - d2#

REM PRINT (v# + (v# - r# + 1)) - r#; (v# + (v# - r# + 2)) - r#; d#; d2#; z;
vt#; rt#; vt# - rt#

PRINT (v# + (v# - r# + 1)) - r#; (v# + (v# - r# + 2)) - r#; "|"; d#; d2#; "|";
z; vt# - rt#; vt# - stir1#; rt# - stir2#

stir1# = vt#

stir2# = rt#

a$ = INKEY$

REM INPUT a$

IF a$ = "s" THEN STOP

d# = d# + 2

d2# = d2# + 2

LOOP
```

```
"C:\Users\Reactor1967\vcns\work\062803B.BAS"

DIM can(1000)

r# = 1

dist# = 0

CLS

count1 = 1

count2 = 1

DO

v# = r# + dist#

PRINT ; dist#; v# + (v# - r# + 1); v# + (v# - r# + 2)

can(count1) = (v# + (v# - r# + 1)) - r#

count1 = count1 + 1

can(count1) = (v# + (v# - r# + 2)) - r#

count1 = count1 + 1

dist# = can(count2)

count2 = count2 + 1

IF dist# >= 99 THEN

INPUT a$

IF a$ = "s" THEN STOP

LOOP
```

```
"C:\Users\Reactor1967\vcns\work\062803A.BAS"

v# = 1

r# = 1

CLS

RANDOMIZE TIMER

DO

z = INT(RND * 2) + 0

a# = v# - r#

IF z = 0 THEN v# = v# + (v# - r#) + 1

IF z = 1 THEN v# = v# + (v# - r#) + 2

b# = v# - r#

PRINT , , z; v#; r#; v# - r#

a$ = INKEY$

IF a$ = "s" THEN STOP

test = v# >= (r# + 99)

IF test = -1 THEN r# = r# + 100

z = INT(RND * 2) + 0

c# = v# - r#

IF z = 0 THEN v# = v# + (v# - r#) + 1

IF z = 1 THEN v# = v# + (v# - r#) + 2

d# = v# - r#

PRINT , , z; v#; r#; v# - r#; a#; b#; c#; d#

a$ = INKEY$

IF a$ = "s" THEN STOP
```

```
test = v# >= (r# + 99)
IF test = -1 THEN r# = r# + 100
LOOP
```

```
"C:\Users\Reactor1967\vcns\work\062103C.BAS"
zoo = 1
t$ = "e:\db.txt"
t2$ = "e:\lb.txt"
OPEN t2$ FOR OUTPUT AS #1
CLS
DO
CLOSE #2
OPEN t$ FOR INPUT AS #2
FOR count = 1 TO 401
INPUT #2, z1, z2, z3, z4, z5, z6
test = (z3 = zoo)
IF test = -1 THEN WRITE #1, z1, z2, z3, z4, z5, z6
NEXT count
zoo = zoo + 1
PRINT zoo
a$ = INKEY$
```

```
IF a$ = "s" THEN EXIT DO
IF zoo = 403 THEN EXIT DO
LOOP
CLOSE #1
CLOSE #2
```

```
"C:\Users\Reactor1967\vcns\work\062103B.BAS"
DIM can1(401)
DIM can2(401)
DIM can3(401)
DIM can4(401)
DIM can5(401)
DIM can6(401)
zoo = 1
CLS
v# = 1
r# = 1
t$ = "e:\062103.txt"
t2$ = "e:\db.txt"
DO
z = INT(RND * 2) + 0
IF z = 0 THEN v# = v# + (v# - r# + 1)
```

```
IF z = 1 THEN v# = v# + (v# - r# + 2)

PRINT , , v#; r#; v# - r#; zoo

a$ = INKEY$

IF a$ = "s" THEN STOP

store1 = v# - r#

test = (v# - r#) >= 99

IF test = -1 THEN r# = r# + 100

store2 = v# - r#

z = INT(RND * 2) + 0

IF z = 0 THEN v# = v# + (v# - r# + 1)

IF z = 1 THEN v# = v# + (v# - r# + 2)

PRINT , , v#; r#; v# - r#; zoo

a$ = INKEY$

IF a$ = "s" THEN STOP

store3 = v# - r#

test = (v# - r#) >= 99

IF test = -1 THEN r# = r# + 100

store4 = v# - r#

CLOSE #1

OPEN t$ FOR INPUT AS #1

FOR count = 1 TO 200

INPUT #1, z1, z2, z3

a$ = INKEY$

IF a$ = "s" THEN STOP
```

```
test = (store1 = z1) AND (store2 = z2)

IF test = -1 THEN EXIT FOR

NEXT count

c1 = z3

CLOSE #1

OPEN t$ FOR INPUT AS #1

FOR count = 1 TO 200

INPUT #1, z1, z2, z3

a$ = INKEY$

IF a$ = "s" THEN STOP

test = (store3 = z1) AND (store4 = z2)

IF test = -1 THEN EXIT FOR

NEXT count

c2 = z3

CLOSE #1

FOR count = 1 TO 401

test = (store1 = can1(count)) AND (store2 = can2(count)) AND (c1 =
can3(count)) AND (store3 = can4(count)) AND (store4 = can5(count))
AND (c2 = can6(count))

IF test = -1 THEN EXIT FOR

NEXT count

IF test = 0 THEN can1(zoo) = store1

IF test = 0 THEN can2(zoo) = store2

IF test = 0 THEN can3(zoo) = c1

IF test = 0 THEN can4(zoo) = store3
```

```
IF test = 0 THEN can5(zoo) = store4

IF test = 0 THEN can6(zoo) = c2

IF test = 0 THEN zoo = zoo + 1

IF zoo = 401 THEN EXIT DO

LOOP

OPEN t2$ FOR OUTPUT AS #2

FOR count = 1 TO 401

WRITE #2, can1(count), can2(count), can3(count), can4(count),
can5(count), can6(count)

NEXT count

CLOSE #2

CLOSE #1

PRINT "All Done"
```

```
"C:\Users\Reactor1967\vcns\work\062103A.BAS"

DIM can1(200)

DIM can2(200)

v# = 1

r# = 1

count2 = 1

CLS
```

```
RANDOMIZE TIMER
DO
z = INT(RND * 2) + 0
IF z = 0 THEN v# = v# + (v# - r# + 1)
IF z = 1 THEN v# = v# + (v# - r# + 2)
PRINT , , v#; r#; v# - r#; count2
c1# = (v# - r#)
a$ = INKEY$
IF a$ = "s" THEN STOP
IF a$ = "x" THEN EXIT DO
test = (v# - r#) >= 99
IF test = -1 THEN r# = r# + 100
c2# = (v# - r#)
FOR count = 1 TO 200
test = (c1# = can1(count)) AND (c2# = can2(count))
IF test = -1 THEN EXIT FOR
NEXT count
IF test = 0 THEN can1(count2) = c1#
IF test = 0 THEN can2(count2) = c2#
IF test = 0 THEN count2 = count2 + 1
LOOP
t$ = "e:\062103.txt"
OPEN t$ FOR OUTPUT AS #1
FOR count = 1 TO 200
```

```
WRITE #1, can1(count), can2(count), count

PRINT , , can1(count), can2(count), count

INPUT a$

IF a$ = "s" THEN CLOSE #1

IF a$ = "s" THEN STOP







NEXT count

CLOSE #1
```

```
"C:\Users\Reactor1967\vcns\work\061803A.BAS"

DIM can1(500)

DIM can2(500)

count = 1

v# = 99

r# = 1

CLS

RANDOMIZE TIMER

DO
```

```
z = INT(RND * 2) + 0
IF flag = 1 THEN z = 1
IF z = 0 THEN b = 1
IF z = 1 THEN b = 2
test = (v# + (v# - r# + b)) >= r# + 99
IF z = 0 THEN v# = v# + (v# - r# + 1)
IF z = 1 THEN v# = v# + (v# - r# + 2)
PRINT , , z; v#; r#; v# - r#; count
IF r# > v# THEN STOP
IF v# - r# = 99 THEN flag = 1 ELSE flag = 0
a$ = INKEY$
IF a$ = "s" THEN EXIT DO
psy1 = (v# - r#)
IF test = -1 THEN r# = r# + 100
psy2 = (v# - r#)
IF psy1 = psy2 THEN GOTO skip:
FOR count2 = 1 TO 500
test = (psy1 = can1(count2)) AND (psy2 = can2(count2))
IF test = -1 THEN EXIT FOR
NEXT count2
IF test = 0 THEN can1(count) = psy1
IF test = 0 THEN can2(count) = psy2
IF test = 0 THEN count = count + 1
skip:
```

```
LOOP
FOR count = 1 TO 500
PRINT can1(count); can2(count)
INPUT a$
IF a$ = "s" THEN STOP
NEXT count
```

```
"C:\Users\Reactor1967\vcns\work\061703A.BAS"
REM changing the equations around its possible to control how fast
REM v# - r# increases or what its divisable by.
v# = 1
r# = 1
CLS
RANDOMIZE TIMER
DO
z = INT(RND * 2) + 0
lb1# = v# - r#
REM                    +/1        +/-     I +/1 N
REM IF z = 0 THEN v# = v# + INT((v# - r#) / 2.5) + 1
REM IF z = 1 THEN v# = v# + INT((v# - r#) / 2.5) + 2
IF z = 0 THEN v# = v# + (v# - r#) + 1: REM also can do this when cal. r#
IF z = 1 THEN v# = v# + (v# - r#) + 2: REM also can do this when cal. r#
```

```
m# = (v# / 4) - INT(v# / 4)
PRINT , , z; v#; r#; v# - r#; dist#
sr# = r#
INPUT a$
IF a$ = "s" THEN STOP
IF a$ = "d" THEN GOTO decode:
dist# = v# - r#
IF z = 0 THEN dist# = dist# - 1
IF z = 1 THEN dist# = dist# - 2
dist# = dist# / 2
dist# = dist# + 1
r# = v# - dist#
LOOP
decode:
DO
dist# = v# - r#
dist# = INT(dist# / 2)
v# = v# - dist#
PRINT , , v#; r#; v# - r#
INPUT a$
IF a$ = "s" THEN STOP
IF a$ = "-" THEN r# = r# - 4
LOOP
```

```
"C:\Users\Reactor1967\vcns\work\061603B.BAS"

DIM a(1000)

DIM b(1000)

count = 1

v# = 1

r# = 1

CLS

RANDOMIZE TIMER

DO

z = INT(RND * 2) + 0

IF z = 0 THEN v# = v# + (v# - r# + 1)

IF z = 1 THEN v# = v# + (v# - r# + 2)

PRINT , , z; v#; r#; v# - r#; count

a$ = INKEY$

IF a$ = "s" THEN EXIT DO

IF a$ = "s" THEN STOP

c = v# - r#

test = (v# >= (r# + 99))

IF test = -1 THEN r# = r# + 100

d = v# - r#

FOR count2 = 1 TO 1000

test = (c = a(count2)) AND (d = b(count2))
```

```
IF test = -1 THEN EXIT FOR

NEXT count2

IF test = 0 THEN a(count) = c

IF test = 0 THEN b(count) = d

IF test = 0 THEN count = count + 1

LOOP

FOR count = 1 TO 1000

PRINT a(count); b(count)

INPUT a$

IF a$ = "s" THEN EXIT FOR

NEXT count
```

"C:\Users\Reactor1967\vcns\work\061603A.BAS"

```
REM ALMOST THERE IT SEEMS!!!. Alright, given a choice of two distances to

REM code to so that we always now if we know N we know which distance to

REM decode with when we repeat our decode sequence. Know, lets do the same

REM but with a twist. Keep R odd always so when we decode we know our N

REM value.

DIM vr(1000)
```

```
CLS
RANDOMIZE TIMER
count = 1
DO
z = INT(RND * 2) + 0
IF z = 0 THEN v# = v# + (v# - r# + 1)
IF z = 1 THEN v# = v# + (v# - r# + 2)
test = 0
FOR count2 = 1 TO 1000
test = ((v# - r#) = vr(count2))
IF test = -1 THEN EXIT FOR
NEXT count2
IF test = 0 THEN vr(count) = (v# - r#)
IF test = 0 THEN count = count + 1
PRINT , , v#; r#; v# - r#; count
a$ = INKEY$
IF a$ = "s" THEN GOTO ot:
test = (v# >= (r# + 99))
IF test = -1 THEN r# = r# + 100
LOOP UNTIL count = 200
ot:
t$ = "c:\061603a.txt"
OPEN t$ FOR OUTPUT AS #1
FOR count = 1 TO 1000
```

```
IF vr(count) > 0 THEN WRITE #1, vr(count)

NEXT count

WRITE #1, 99999

CLOSE #1

PRINT "Data Base initialized"

INPUT z$

IF z$ = "n" THEN STOP

v# = 1000

redo2:

OPEN t$ FOR INPUT AS #1

DO

z = INT(RND * 2) + 0

retry:

DO

INPUT #1, a#

t1 = (a# / 2) - INT(a# / 2) = .5

t2 = (a# / 2) - INT(a# / 2) = 0

t3 = (z = 1) AND (t2 = -1)

t4 = (z = 0) AND (t1 = -1)

t5 = (t3 = -1) OR (t4 = -1)

IF t5 = -1 THEN EXIT DO

IF a# = 99999 THEN EXIT DO

a$ = INKEY$

IF a$ = "s" THEN STOP
```

```
LOOP
IF a# = 99999 THEN EXIT DO
b# = a#
IF (b# / 2) - INT(b# / 2) = .5 THEN b# = b# - 1
b# = b# / 2
b# = b# + 1
v# = v# + b#
r# = v# - a#
IF (r# / 2) - INT(r# / 2) = 0 THEN GOTO retry:
PRINT , , z; v#; r#; v# - r#; r# - store#
store# = r#
IF r# > v# THEN STOP
IF (v# - r#) <> a# THEN PRINT "task failed"
IF (v# - r#) <> a# THEN STOP
a$ = INKEY$
IF a$ = "s" THEN STOP
LOOP
CLOSE #1
GOTO redo2:
```

"C:\Users\Reactor1967\vcns\work\061503A.BAS"

```
REM Killing a fly with the cannon ball approach. Control the distance
```

REM you code to between v# - r#. Be able to code this distance going up

REM or going down to specific even or odd distances. When you decode you

REM have a choice of two distances you will code to. Look at your v#

REM to see if its even or odd then you know the distance between v# - r#

REM repeat as needed. This is still in development.

```
RANDOMIZE TIMER

v# = 1001

dist# = 0

CLS

redo:

DO

dist# = INT(RND * 292) + 2

z = INT(RND * 2) + 0

test = ((dist# / 2) - INT(dist# / 2) = .5) AND z = 0

IF test = 0 THEN dist# = dist# - 1

REM dist# = dist# + 101

b# = dist#

IF (dist# / 2) - INT(dist# / 2) = .5 THEN dist# = dist# - 1

dist# = dist# / 2

dist# = dist# + 1

v# = v# + dist#

r# = v# - b#

PRINT , , z; v#; r#; v# - r#; b#
```

```
store# = r#

IF (v# - r#) <> b# THEN PRINT , , "Task Failed"

IF (v# - r#) <> b# THEN STOP

a$ = INKEY$

REM INPUT a$

IF a$ = "s" THEN STOP

IF v# >= 99999 THEN EXIT DO

LOOP

DO

dist# = INT(RND * 292) + 2

z = INT(RND * 2) + 0

test = ((dist# / 2) - INT(dist# / 2) = .5) AND z = 0

IF test = 0 THEN dist# = dist# - 1

REM dist# = dist# + 101

b# = dist#

IF (dist# / 2) - INT(dist# / 2) = .5 THEN dist# = dist# - 1

dist# = dist# / 2

dist# = dist# + 1

v# = v# - dist#

r# = v# + b#

PRINT , , z; v#; r#; r# - v#; b#

store# = r#

IF (r# - v#) <> b# THEN PRINT , , "Task Failed"

IF (r# - v#) <> b# THEN STOP
```

```
a$ = INKEY$
REM INPUT a$
IF a$ = "s" THEN STOP
IF v# <= 10100 THEN EXIT DO
LOOP
GOTO redo:
```

```
"C:\Users\Reactor1967\vcns\work\61403A.BAS"
spr# = (v# - r#) - spv# + N
REM
v# = 1
r# = 1
less# = 9999
great# = 0
CLS
RANDOMIZE TIMER
numberbase = 2
DO
z = INT(RND * 2) + 0
IF z = 0 THEN N = 1
IF z = 1 THEN N = 2
v# = v# + ((v# - r#) * (numberbase - 1)) + N
```

```
PRINT , "N"; z; v#; r#; "v# - r# "; v# - r#; "spvr "; (v# - r#) - store#; "spr
"; r# - store2#; "spv "; v# - store3#

store# = (v# - r#)

store2# = r#

store3# = v#

a$ = INKEY$

INPUT a$

IF a$ = "s" THEN STOP

IF v# >= (r# + 99) THEN r# = r# + 100

LOOP
```

```
"C:\Users\Reactor1967\vcns\work\061403C.BAS"

REM Being able to control the speed of v#, the speed of r#, and the speed

REM of v# - r# is very very inportant. To be able to control how these

REM variables go up and down seems the key here. So to had to another

REM peice of the puzzle here is a new equation.

REM the speed of (v# - r#) = (v# - r#) - (speed of r#) + (speed of v#)

v# = 49

r# = 1

z = 0

CLS

RANDOMIZE TIMER
```

```
numberbase = 2

DO

store# = v#

v# = v# + ((v# - r#) * (numberbase - 1)) + z

PRINT z; v#; r#; v# - r#; v# - store1#; r# - store2#; v# - r# - store3#

store1# = v#

store2# = r#

store3# = v# - r#

a$ = INKEY$

IF a$ = "s" THEN STOP

REM r# = r# + ((v# - r#) - (v# - store#) + z)

z = INT(RND * 2) + 1

r# = r# + ((v# - r#) - 100 + z)

IF (r# / 2) - INT(r# / 2) = 0 THEN r# = r# - 1

LOOP
```

```
"C:\Users\Reactor1967\vcns\work\061403B.BAS"

vr# = 0

n1 = 0

n2 = 1

spv = 100
```

```
CLS
DO
PRINT vr#; vr# - spv + n1; vr# - spv + n2
INPUT a$
IF a$ = "s" THEN STOP
vr# = vr# + 1
LOOP
```

```
"C:\Users\Reactor1967\vcns\work\061403A.BAS"
RANDOMIZE TIMER
v# = 5000
r# = 4901
vr# = v# - r#
spr# = 100
z = 0
DO
sz = z
store1# = vr#
z = INT(RND * 2) + 0
IF z = sz THEN vr# = vr# + 100 ELSE vr# = vr# + 99
b# = vr#
IF (b# / 2) - INT(b# / 2) = .5 THEN b# = b# - 1
```

```
b# = b# / 2

b# = b# + 1

sr# = store1# - b#

r# = r# + sr#

v# = v# + b#

vr# = v# - r#

PRINT , , z; v#; r#; vr#; b#; sr#; (v# - r#) - store2#

store2# = v# - r#

INPUT a$

IF a$ = "s" THEN STOP

LOOP
```

```
"C:\Users\Reactor1967\vcns\work\61303A.BAS"

t1# = 0

t2# = 0

CLS

count = 1

DO

REM spr = speed of r#

REM spv = speed of v#

PRINT "spr at 0"; t1# - t2# + 1; "spr at 1"; t1# - t2# + 2; "v# - r#"; t1#;
"spv"; t2#
```

```
count = count + 1
t1# = t1# + 1
IF count = 3 THEN t2# = t2# + 1
IF count = 3 THEN count = 1
IF t2# = 101 THEN t2# = 1
a$ = INKEY$
REM INPUT a$
IF a$ = "s" THEN STOP
LOOP
```

```
"C:\Users\Reactor1967\vcns\work\61103E.BAS"
DIM g(250)
count = 1
v# = 1
r# = 1
CLS
RANDOMIZE TIMER
ON ERROR GOTO storeit:
redo:
DO
z = INT(RND * 2) + 0
IF z = 0 THEN v# = v# + (v# - r# + 1)
```

```
IF z = 1 THEN v# = v# + (v# - r# + 2)

PRINT , , z; v#; r#; v# - r#; count; v# - store#

a$ = INKEY$

IF a$ = "s" THEN GOTO storeit:

test = v# >= (r# + 99)

IF test = -1 THEN EXIT DO

store# = v#

LOOP

test = (v# - store#)

FOR dad = 1 TO 250

test2 = (test = g(dad))

IF test2 = -1 THEN EXIT FOR

a$ = INKEY$

IF a$ = "s" THEN GOTO storeit:

NEXT dad

IF test2 = 0 THEN g(count) = v# - store#

IF test2 = 0 THEN count = count + 1

a$ = INKEY$

IF a$ = "s" THEN GOTO storeit:

r# = r# + 100

GOTO redo:

storeit:

t$ = "c:\db.txt"

OPEN t$ FOR OUTPUT AS #1
```

```
FOR cc = 1 TO 250
dt = g(cc)
IF dt > 0 THEN WRITE #1, dt
NEXT cc
WRITE #1, v#, r#
CLOSE #1
PRINT "Your done"
```

```
"C:\Users\Reactor1967\vcns\work\61103D.BAS"
v# = 127
r# = 1
CLS
RANDOMIZE TIMER
redo:
DO
d# = (v# - r#)
IF (d# / 2) - INT(d# / 2) = .5 THEN d# = d# - 1
d# = d# / 2
z = INT(RND * 2) + 0
IF z = 0 THEN v# = v# - (v# - d#) - 1
IF z = 1 THEN v# = v# - (v# - d#) - 2
PRINT , , z; v#; r#; v# - r#
```

```
a$ = INKEY$

INPUT a$

IF a$ = "s" THEN STOP

d# = v# - r#

IF (d# / 2) - INT(d# / 2) = .5 THEN d# = d# - 1

d# = d# / 2

IF v# - (v# - d#) - 2 < 1 THEN EXIT DO

LOOP

DO

z = INT(RND * 2) + 0

IF z = 0 THEN v# = v# + (v# - r#) + 1

IF z = 1 THEN v# = v# + (v# - r#) + 2

PRINT , , z; v#; r#; v# - r#

a$ = INKEY$

REM INPUT a$

IF a$ = "s" THEN STOP

IF v# > 100 THEN EXIT DO

LOOP

GOTO redo:
```

```
"C:\Users\Reactor1967\vcns\work\61103B.BAS"

RANDOMIZE TIMER
```

```
CLS
v# = 1
r# = 1
redo:
DO
z = INT(RND * 2) + 0
IF z = 0 THEN v# = v# + (v# - r# + 1)
IF z = 1 THEN v# = v# + (v# - r# + 2)
PRINT , , z; v#; r#; v# - r#
a$ = INKEY$
REM INPUT a$
IF a$ = "s" THEN STOP
test = v# >= (r# + 99)
LOOP UNTIL test = -1
r# = v#
DO
test = (v# <= (r# - 9))
IF test = -1 THEN r# = r# - 10
IF test = -1 THEN z = 1 ELSE z = 0
IF z = 0 THEN v# = v# + (v# - r#) - 1
IF z = 1 THEN v# = v# + (v# - r#) - 2
PRINT , , z; v#; r#; r# - v#
a$ = INKEY$
REM INPUT a$
```

```
IF a$ = "s" THEN STOP

LOOP UNTIL v# <= 10

r# = 1

GOTO redo:
```

```
"C:\Users\Reactor1967\vcns\work\61103.BAS"

v# = 1

r# = 1

CLS

g# = v#

t$ = "c:\db.txt"

CLOSE #1

OPEN t$ FOR OUTPUT AS #1

redo:

DO

v# = v# + (v# - r#) + 1

PRINT , , v#; r#; v# - r#

b$ = LTRIM$(b$)

b$ = RTRIM$(b$)

b$ = b$ + "0"

a$ = INKEY$

REM INPUT a$
```

```
IF a$ = "s" THEN STOP
test = (v# >= (r# + 99))
IF test = -1 THEN EXIT DO
LOOP
WRITE #1, g#, b$, v# - r#, (v# - r#) - 100
v# = g#
b$ = ""
DO
v# = v# + (v# - r# + 2)
PRINT , , v#; r#; v# - r#
b$ = LTRIM$(b$)
b$ = RTRIM$(b$)
b$ = b$ + "1"
a$ = INKEY$
IF a$ = "s" THEN STOP
test = (v# >= (r# + 99))
IF test = -1 THEN EXIT DO
LOOP
WRITE #1, g#, b$, v# - r#, (v# - r#) - 100
b$ = ""
g# = g# + 1
v# = g#
IF g# > 99 THEN CLOSE #1
IF g# > 99 THEN STOP
```

GOTO redo:

"C:\Users\Reactor1967\vcns\work\061103LB.BAS"

REM here look at your speed of v# and goto a data base to

REM find out what your speed of r# is. I have to create that

REM data base but that can be done with this program. Store

REM all specific speeds of v# with there speeds of r#

REM so that v# = v# + (v# + x#) + N so that the speed of v# has a specific

REM speed of r#. Calculate this and test before you do the real code.

REM you will have to test specific values of n each time before you code

REM until you find a match in your data base.

REM your N value will have to be calculated from your current v# - r#

REM use specific N values for each current specific v# - r#. Doing this

REM I believe will solve all the problems. We wil see

REM speed of r# = ((v# - r#) - (speed of v#)) + N value of next code

v# = 900

RANDOMIZE TIMER

CLS

DO

z = INT(RND * 2) + 0

IF (v# / 2) - INT(v# / 2) = .5 THEN r# = v# - 100

```
IF (v# / 2) - INT(v# / 2) = 0 THEN r# = r# - 99

IF z = 0 THEN v# = v# + (v# - r#) + 1

IF z = 1 THEN v# = v# + (v# - r#) + 2

PRINT , , z; v#; r#; v# - r#; r# - store#; v# - store2#

REM IF v# - store2# = 102 THEN INPUT z$

IF z$ = "s" THEN STOP

store# = r#

store2# = v#

a$ = INKEY$

IF a$ = "s" THEN STOP

LOOP
```

```
"C:\Users\Reactor1967\vcns\work\061103C.BAS"

t1$ = "c:\db.txt"

t2$ = "c:\db2.txt"

CLS

OPEN t1$ FOR INPUT AS #1

OPEN t2$ FOR OUTPUT AS #2

v# = 128

r# = 197

DO

INPUT #1, z, b1$, z2
```

```
IF z = 99 THEN GOTO ot:
REM PRINT z, b1$, z2
g = LEN(b1$)
FOR count = 1 TO g
PRINT , , z; v#; r#; v# - 1
z = VAL(MID$(b1$, g, 1))
IF z = 0 THEN v# = v# - (r# - v#) + 1
IF z = 1 THEN v# = v# - (r# - v#) + 2
REM IF v# - r# <= 47 THEN EXIT DO
NEXT count
PRINT z, b1$, z2

CLOSE
STOP
LOOP
ot:
CLOSE
```

```
"C:\Users\Reactor1967\vcns\work\61003A.BAS"
CLS
t$ = "c:\db.txt"
t1$ = "c\db2.txt"
```

```
OPEN t$ FOR OUTPUT AS #1
count = 0
CLS
DO
INPUT z1, z2, b$, z3, z4


SUB bn (count, bin$)
IF count < 0 THEN STOP
IF count > 9999999 THEN STOP
store = count
d# = 1
count2 = 1
DO
IF d# >= count THEN EXIT DO
d# = d# * 2
count2 = count2 + 1
LOOP
IF d# > count THEN count2 = count2 - 1
count2 = count2 - 1
bin$ = ""
DO
a# = (2 ^ count2)
IF a# <= count THEN bin$ = bin$ + "1"
IF a# > count THEN bin$ = bin$ + "0"
```

```
IF a# <= count THEN count = count - a#

count2 = count2 - 1

IF count2 < 0 THEN EXIT DO

LOOP

count = store

END SUB
```

```
"C:\Users\Reactor1967\vcns\work\61003.BAS"

REM this program creates a database that vcns can use to function with.

REM the purpose of this data base is to find stable sequences of patterns

REM of numbers that can be run and used to represent 1's and 0's.

REM sequences may be needed to be used as switches to switch from
different

REM patterns. Anyway if a specific pattern can be establish that can
switch

REM back and forth from each other and predictable enough to reconize
thus

REM decode with then it might be possible to code binary data with it.

DECLARE SUB bn (count!, bin$)

CLS

t$ = "c:\db.txt"

OPEN t$ FOR OUTPUT AS #1

v# = 999
```

```
r# = 999

store# = v#

count = 0

LB# = v# - r#

DO

CALL bn(count, bin$)

binstore = count

g = LEN(bin$)

flag = 0

realdeal$ = ""

DO

PRINT , , v#; r#; v# - r#; z

IF g > 0 THEN z = VAL(MID$(bin$, g, 1)) ELSE z = 0

IF z = 0 THEN flag = 1

IF z = 0 THEN realdeal$ = realdeal$ + "0"

IF z = 1 THEN realdeal$ = realdeal$ + "1"

IF z = 0 THEN v# = v# + (v# - r#) + 1

IF z = 1 THEN v# = v# + (v# - r#) + 2

g = g - 1

test = (v# >= (r# + 99))

IF test = -1 THEN EXIT DO

a$ = INKEY$

IF a$ = "s" THEN STOP

LOOP
```

```
PRINT , , v#; r#; v# - r#; z; v# - (r# - 100)
PRINT , , LB#; bin$; (v# - r#); binstore; realdeal$
a$ = INKEY$
REM INPUT a$
IF a$ = "s" THEN STOP
crank# = v# - r#
WRITE #1, (100 + LB#), LB#, realdeal$, crank#, v# - (r# + 100)
count = count + 1
IF flag = 0 THEN count = 0
IF flag = 0 THEN store# = store# + 1
PRINT "                    "
v# = store#
LB# = v# - r#
IF v# - r# >= 99 THEN WRITE #1, 999, "999", 999
IF v# - r# >= 99 THEN CLOSE #1
IF v# - r# >= 99 THEN STOP
LOOP


SUB bn (count, bin$)
IF count < 0 THEN STOP
IF count > 9999999 THEN STOP
store = count
d# = 1
count2 = 1
```

```
DO

IF d# >= count THEN EXIT DO

d# = d# * 2

count2 = count2 + 1

LOOP

IF d# > count THEN count2 = count2 - 1

count2 = count2 - 1

bin$ = ""

DO

a# = (2 ^ count2)

IF a# <= count THEN bin$ = bin$ + "1"

IF a# > count THEN bin$ = bin$ + "0"

IF a# <= count THEN count = count - a#

count2 = count2 - 1

IF count2 < 0 THEN EXIT DO

LOOP

count = store

END SUB
```

```
"C:\Users\Reactor1967\vcns\work\061003A.BAS"

CLS

t$ = "c:\lb.txt"
```

```
OPEN t$ FOR OUTPUT AS #1
v# = 1
r# = 1
g# = v#
redo:
z = 1
h = (v# - r#)
DO
PRINT , , z; v#; r#; v# - r#
b$ = b$ + STR$(z)
b$ = LTRIM$(b$)
b$ = RTRIM$(b$)
IF z = 0 THEN v# = v# + (v# - r#) + 1
IF z = 1 THEN v# = v# + (v# - r#) + 2
test = (v# >= (r# + 99))
IF test = -1 THEN EXIT DO
a$ = INKEY$
IF a$ = "s" THEN STOP
z = 0
LOOP
PRINT , , z; v#; r#; v# - r#
WRITE #1, h, b$, (v# - r#), (v# - r#) - 100
g# = g# + 1
v# = g#
```

```
IF v# - r# >= 102 THEN CLOSE #1

IF v# - r# >= 102 THEN STOP

b$ = ""

GOTO redo:
```

```
"C:\Users\Reactor1967\vcns\work\060603A.BAS"

REM randomly using sequences to represent binary data.

REM Each sequence has a present v# - r# that it came from

REM and goes to. Im trying to use this as a focal point to

REM tie coding random binary data.

t$ = "c:\db.txt"

CLS

v# = 1

r# = 1

RANDOMIZE TIMER

DO

z = INT(RND * 2) + 0

CLOSE #1

OPEN t$ FOR INPUT AS #1

DO

INPUT #1, z1, z2, a$, z3, z4

test = ((v# - r#) = z2) AND (z = VAL(MID$(a$, 1, 1)))
```

```
IF (v# - r#) = 49 THEN test = -1

IF test = -1 THEN EXIT DO

a$ = INKEY$

IF a$ = "s" THEN STOP

IF z1 = 999 THEN EXIT DO

LOOP

PRINT , , "----------------------------------------"

FOR count = 1 TO LEN(a$)

z = VAL(MID$(a$, count, 1))

IF z = 0 THEN v# = v# + (v# - r#) + 1

IF z = 1 THEN v# = v# + (v# - r#) + 2

PRINT , , z1; z2; v#; r#; v# - r#; z3; z4

a$ = INKEY$

IF a$ = "s" THEN STOP

NEXT count

r# = v# - z4

a$ = INKEY$

IF a$ = "s" THEN STOP


LOOP
```

```
"C:\Users\Reactor1967\vcns\work\060203A.BAS"

CLS

RANDOMIZE TIMER

redo:

v# = 1

r# = 1

z = 1

CLS

DO

IF z = 0 THEN v# = v# + (v# - r# + 1)

IF z = 1 THEN v# = v# + (v# - r# + 2)

PRINT , , "+"; z; v#; r#; v# - r#

a$ = INKEY$

IF a$ = "s" THEN t$ = "c:\ps.txt"

IF a$ = "s" THEN fh = 1 ELSE fh = 0

IF fh = 1 THEN CLOSE

IF fh = 1 THEN OPEN t$ FOR OUTPUT AS #1

IF fh = 1 THEN WRITE #1, "+", v#, r#, test

IF fh = 1 THEN CLOSE

IF a$ = "s" THEN STOP

IF a$ = "d" THEN EXIT DO

IF v# >= (r# + 99) THEN z = 1 ELSE z = 0

IF v# >= (r# + 99) THEN r# = r# + 100

LOOP
```

```
test = 0
DO
z2 = (v# / 2) - INT(v# / 2)
IF z2 = .5 THEN z2 = 1
PRINT , , "-"; z2; v#; r#; v# - r#
IF v# <= 1 THEN EXIT DO
test = (v# / 2) - INT(v# / 2) = .5
a$ = INKEY$
REM INPUT a$
IF a$ = "s" THEN t$ = "c:\ps.txt"
IF a$ = "s" THEN fh = 1 ELSE fh = 0
IF fh = 1 THEN CLOSE
IF fh = 1 THEN OPEN t$ FOR OUTPUT AS #1
IF fh = 1 THEN WRITE #1, "-", v#, r#, test
IF fh = 1 THEN CLOSE
IF a$ = "s" THEN STOP
d# = v# - r#
IF (d# / 2) - INT(d# / 2) = .5 THEN d# = d# - 1
d# = d# / 2
d# = d# + 1
v# = v# - d#
IF v# <= 3 THEN test = 0
IF test = -1 THEN r# = r# - 100
LOOP
```

```
GOTO redo:
```

```
"C:\Users\Reactor1967\vcns\work\060103.BAS"
REM its possible using a preset sequence of 1's and 0's with a
REM preset sequence of increasing r to code a random pattern that
REM will stay in line and be decodable. When you decode you read the
REM sequence and know if you have a 1 or a 0 and know how much to
REM decrease r by.
r# = 1
v# = 1
CLS
redo:
RANDOMIZE TIMER
FOR count = 1 TO 7
v# = v# + (v# - r# + 1)
PRINT v#; r#; v# - r#
a$ = INKEY$
IF a$ = "s" THEN STOP
NEXT count
r# = r# + 126
FOR count = 1 TO 7
v# = v# + (v# - r# + 2)
```

```
PRINT v#; r#; v# - r#
a$ = INKEY$
IF a$ = "s" THEN STOP
NEXT count
r# = r# + 126
INPUT a$
IF a$ = "s" THEN STOP
GOTO redo:
```

```
"C:\Users\Reactor1967\vcns\work\053003A.BAS"
REM it seems like it does not matter what is bigger or smaller v or r as
REM long as they stay within range of each other and it can be decoded.
v# = 1
r# = 1
RANDOMIZE TIMER
CLS
DO
z = INT(RND * 2) + 0
IF z = 0 THEN v# = v# + ABS(v# - r#) + 1
IF z = 1 THEN v# = v# + ABS(v# - r#) + 2
PRINT , , z; v#; r#; v# - r#
IF v# < r# THEN STOP
```

```
a$ = INKEY$
REM INPUT a$
IF a$ = "s" THEN STOP
IF v# >= (r# + 99) THEN flag = 1 ELSE flag = 0
IF flag = 1 THEN r# = r# + 100
LOOP
```

"C:\Users\Reactor1967\vcns\work\053003.BAS"

```
count = 1
CLS
DO
d# = 1
count2 = 1
DO
IF d# >= count THEN EXIT DO
d# = d# * 2
count2 = count2 + 1
a$ = INKEY$
IF a$ = "s" THEN STOP
LOOP
IF d# > count THEN count2 = count2 - 1
PRINT , count; count2; ,
```

```
REM PRINT , , ,
store = count
count2 = count2 - 1
bin$ = ""
DO
a# = (2 ^ count2)
REM IF a# <= count THEN PRINT "1";
IF a# <= count THEN bin$ = bin$ + "1"
REM IF a# > count THEN PRINT "0";
IF a# > count THEN bin$ = bin$ + "0"
IF a# <= count THEN count = count - a#
count2 = count2 - 1
IF count2 < 0 THEN EXIT DO
a$ = INKEY$
IF a$ = "s" THEN STOP
LOOP
count = store
PRINT bin$
a$ = INKEY$
INPUT a$
IF a$ = "s" THEN STOP
count = count + 1
LOOP
```

```
"C:\Users\Reactor1967\vcns\work\052703A.BAS"

REM This program confirms that the starting distance and ending

REM distance is the same for each binary pattern of 1's and 0's.

REM using this a database can be constructed to make the coding and

REM decoding of binary data possible for increasing r# and keeping r
and

REM v in a specific range.

CLS

v# = 350

INPUT "Enter distance to r#"; d#

r# = v# - d#

DO

INPUT z

IF z > 1 THEN EXIT DO

IF z = 0 THEN v# = v# + (v# - r# + 1)

IF z = 1 THEN v# = v# + (v# - r# + 2)

PRINT , , z; v#; r#; v# - r#

IF v# - r# >= 100 THEN EXIT DO

LOOP
```

"C:\Users\Reactor1967\vcns\work\042703A.BAS"

```
REM use this to get sv# and sr# but us sv# to derive distances

REM make a chart.

d# = 0

sv# = 0

sr# = 0

CLS

z = 0

DO

d# = d# + 1

sv# = INT(d# / 2)

sr# = d# - sv#

IF (sr# / 2) - INT(sr# / 2) = .5 THEN flag = 1 ELSE flag = 0

IF flag = 1 THEN sv# = sv# + 1

IF flag = 1 THEN sr# = sr# - 1

PRINT , , z; d#; sv#; sr#

INPUT a$

IF a$ = "s" THEN STOP

IF z = 0 THEN z = 1 ELSE z = 0

LOOP
```

"C:\Users\Reactor1967\vcns\work\042003A.BAS"

```
v# = 2
r# = 1
CLS
RANDOMIZE TIMER
DO
z = INT(RND * 2) + 0
IF z = 0 THEN v# = v# + (v# - r# + 1)
IF z = 1 THEN v# = v# + (v# - r# + 2)
PRINT z; v#; r#; v# - r#; r# - sd#
sd# = r#
a# = (v# - r#)
test = (a# / 2) - INT(a# / 2) = 0
IF test = -1 THEN a# = (a# / 2)
IF test = 0 THEN a# = (a# / 2) - .5
r# = r# + a#
a$ = INKEY$
IF a$ = "s" THEN STOP
LOOP
```

```
"C:\Users\Reactor1967\vcns\work\041903A.BAS"
REM coding and incoding chart
```

```
REM c1# - i1# is what r# increases by for the dist# c1

REM c2# - i2# is what r# increasey by for the dist c2#

REM when decoding if your distance is o3# or o4# r# = r# - (i2# - c2#)

REM when decoding if your distance is o1# or o2# then r# = r# - (i1# -
c1#)

c# = 2

CLS

DO

i1# = (c# / 2)

o1# = ((c# - i1#) * 2) + 1

o2# = ((c# - i1#) * 2) + 2

c1# = c#

c# = c# + 1

c2# = c#

i2# = (c# / 2) - .5

o3# = ((c# - i2#) * 2) + 1

o4# = ((c# - i2#) * 2) + 2

PRINT c1#; c2#; ; c1# - i1#; c2# - i2#; ; o1#; o2#; ; o3#; o4#

a$ = INKEY$

REM INPUT a$

IF a$ = "s" THEN STOP

c# = c# + 1

LOOP
```

```
"C:\Users\Reactor1967\vcns\work\041303B.BAS"
REM see if can use this as a decode map.
REM can look at v# and r# when there are two
REM decode distances to tell which one is correct.
REM this is true because r# here will always be odd
REM if starting with a odd r or even if starting with
REM a even r. Looking at the previous v# will tell
REM if had a previous odd(v# - r) or even(v# - r#).
CLS
a# = 1
b# = a#
IF (b# / 2) - INT(b# / 2) = .5 THEN b# = b# - 1
b# = b# / 2
c# = ((a# - b#) * 2) + 1
d# = ((a# - b#) * 2) + 2
DO
IF (b# / 2) - INT(b# / 2) = .5 THEN b# = b# - 1
PRINT , , a#; b#; c#; d#
a$ = INKEY$
INPUT a$
IF a$ = "s" THEN STOP
a# = a# + 1
```

```
b# = a#
IF (b# / 2) - INT(b# / 2) = .5 THEN b# = b# - 1
b# = b# / 2
c# = ((a# - b#) * 2) + 1
d# = ((a# - b#) * 2) + 2
LOOP
```

```
"C:\Users\Reactor1967\vcns\work\041303A.BAS"
z2 = 1
v# = 25
r# = 1
CLS
RANDOMIZE TIMER
DO
dist# = v# - r#
IF z2 = 0 THEN dist# = dist# - 1
IF z2 = 1 THEN dist# = dist# - 2
dist# = dist# / 2
IF (dist# / 2) - INT(dist# / 2) = .5 THEN dist# = dist# - 1
r# = r# + dist#
z = INT(RND * 2) + 0
IF z = 0 THEN v# = v# + (v# - r# + 1)
```

```
IF z = 1 THEN v# = v# + (v# - r# + 2)

PRINT , , z; v#; r#; v# - r#; dist#

a$ = INKEY$

IF a$ = "s" THEN STOP

z2 = z

LOOP
```

```
"C:\Users\Reactor1967\vcns\work\041103D.BAS"

REM g2# - ((v# - r#) / 2 - N)  - N = g1#

REM Note if v# - r# is odd subtract 1: if g2# - N is odd subtract 1

v# = 2

r# = 1

CLS

RANDOMIZE TIMER

dist# = 1

g# = 0

DO

z = INT(RND * 2) + 0

IF z = 0 THEN v# = v# + (v# - r# + 1)

IF z = 1 THEN v# = v# + (v# - r# + 2)

dist# = v# - r#

IF (dist# / 2) - INT(dist# / 2) = .5 THEN dist# = dist# - 1
```

```
dist# = dist# / 2

IF (dist# / 2) - INT(dist# / 2) = .5 THEN dist# = dist# - 1

PRINT z; v#; r#; v# - r#; dist#; r# - g#

REM ------------------------------------------

test1# = (v# - r#)

IF (test1# / 2) - INT(test1# / 2) = .5 THEN test1# = test1# - 1

test1# = test1# / 2

IF z = 0 THEN test1# = test1# - 1

IF z = 1 THEN test1# = test1# - 2

IF (test1# / 2) - INT(test1# / 2) = .5 THEN test1# = test1# - 1

IF test1# <> (r# - g#) THEN PRINT "                              "

REM ------------------------------------------

g# = r#

a$ = INKEY$

REM INPUT a$

IF a$ = "s" THEN STOP

r# = r# + dist#

LOOP
```

"C:\Users\Reactor1967\vcns\work\041103C.BAS"

```
REM g2# - ((v# - r#) / 2 - N)  - N = g1#

REM Note if v# - r# is odd subtract 1: if g2# - N is odd subtract 1

v# = 2

r# = 1

CLS

RANDOMIZE TIMER

dist# = 1

g# = 0

DO

z = INT(RND * 2) + 0

IF z = 0 THEN v# = v# + (v# - r# + 1)

IF z = 1 THEN v# = v# + (v# - r# + 2)

PRINT z; v#; r#; v# - r#; g#;

dist# = v# - r#

a$ = INKEY$

INPUT a$

IF a$ = "s" THEN STOP

r# = r# + g#

IF (dist# / 2) - INT(dist# / 2) = .5 THEN dist# = dist# - 1

dist# = dist# / 2

IF (dist# / 2) - INT(dist# / 2) = .5 THEN dist# = dist# - 1

IF dist# >= 1 THEN g# = dist#

LOOP
```

```
"C:\Users\Reactor1967\vcns\work\041103B.BAS"

REM g2# - ((v# - r#) / 2 - N)  - N = g1#

REM Note if v# - r# is odd subtract 1: if g2# - N is odd subtract 1

v# = 2

r# = 1

CLS

RANDOMIZE TIMER

dist# = 1

g# = 0

DO

z = INT(RND * 2) + 0

IF z = 0 THEN v# = v# + (v# - r# + 1)

IF z = 1 THEN v# = v# + (v# - r# + 2)

PRINT z; v#; r#; v# - r#; g#; g# - sd#

dist# = v# - r#

a$ = INKEY$

INPUT a$

IF a$ = "s" THEN STOP

r# = r# + g#

IF (dist# / 2) - INT(dist# / 2) = .5 THEN dist# = dist# - 1
```

```
dist# = dist# / 2
IF (dist# / 2) - INT(dist# / 2) = .5 THEN dist# = dist# - 1
IF dist# >= 1 THEN g# = dist#
LOOP
```

```
"C:\Users\Reactor1967\vcns\work\041103A.BAS"
v# = 2
r# = 1
CLS
RANDOMIZE TIMER
dist# = 1
DO
z = INT(RND * 2) + 0
dist# = v# - r#
IF (dist# / 2) - INT(dist# / 2) = .5 THEN dist# = dist# - 1
IF z = 0 THEN v# = v# + (v# - r# + 1)
IF z = 1 THEN v# = v# + (v# - r# + 2)
PRINT , , z; v#; r#; v# - r#; g#; g# - sd#
REM IF v# - r# >= 271 THEN STOP
a$ = INKEY$
```

```
sd# = g#

g# = dist#

r# = r# + dist#

INPUT a$

IF a$ = "s" THEN STOP

LOOP
```

```
"C:\Users\Reactor1967\vcns\work\040603A.BAS"

v# = 1

r# = 1

RANDOMIZE TIMER

CLS

DO

z = INT(RND * 2) + 0

dist# = v# - r#

v# = v# + (v# - r#) + z + 1

PRINT , z; v#; r#; v# - r#; dist#; dist# - sd#

IF (dist# / 2) - INT(dist# / 2) = .5 THEN dist# = dist# - 1

sd# = dist#

a$ = INKEY$
```

```
IF a$ = "s" THEN STOP
r# = r# + dist#
LOOP
```

"C:\Users\Reactor1967\vcns\work\040403A.BAS"

```
v# = 1
r# = 1
CLS
DO
z = INT(RND * 2) + 0
IF z = 0 THEN v# = v# + (v# - r# + 1)
IF z = 1 THEN v# = v# + (v# - r# + 3)
IF z = 3 THEN v# = v# + (v# - r# + 2)
a$ = INKEY$
REM INPUT a$
IF a$ = "s" THEN STOP
PRINT , , z; v#; r#; v# - r#
IF v# - r# >= 53 THEN z = 3
IF z = 3 THEN r# = r# + 105
LOOP
```

```
"C:\Users\Reactor1967\vcns\work\033003B.BAS"
v# = 1
r# = 1
sum# = 0
CLS
RANDOMIZE TIMER
DO
z = INT(RND * 2) + 0
speed# = (v# - r#) + z + 1
IF z = 0 THEN v# = v# + (((v# - r#) * 1) + 1)
IF z = 1 THEN v# = v# + (((v# - r#) * 1) + 2)
PRINT , , z; v#; r#; v# - r#; speed#
a$ = INKEY$
REM INPUT a$
IF a$ = "s" THEN STOP
sum# = sum# + speed#
IF sum# >= 100 THEN flag = 1 ELSE flag = 0
IF flag = 1 THEN r# = r# + speed#
IF flag = 1 THEN sum# = 0
LOOP
```

```
"C:\Users\Reactor1967\vcns\work\033003A.BAS"

v# = 20

r# = 1

ex# = 1

speed# = 1

CLS

RANDOMIZE TIMER

DO

z = INT(RND * 2) + 0

speed# = (v# - r#) + z + 1

IF z = 0 THEN v# = v# + (((v# - r#) * 1) + 1)

IF z = 1 THEN v# = v# + (((v# - r#) * 1) + 2)

sum# = sum# + speed#

PRINT , , z; v#; r#; speed#; flag; sum#

sv# = v#

a$ = INKEY$

REM INPUT a$

IF a$ = "s" THEN STOP

IF (v# - r#) >= 99 THEN flag = 1 ELSE flag = 0

IF (v# - r#) >= 99 THEN r# = r# + 100

IF flag = 1 THEN sum# = 0

LOOP
```

```
"C:\Users\Reactor1967\vcns\work\032803A.BAS"
v# = 1
r# = 1
store# = 1
CLS
RANDOMIZE TIMER
DO
REM use the equation for a trangle here to experiment with
z = INT(RND * 2) + 0
IF z = 0 THEN v# = v# + (v# - r# + 1)
IF z = 1 THEN v# = v# + (v# - r# + 2)
a# = v# - r#
b# = store# - r#
c# = v# - store#
PRINT , , z; v#; r#; v# - r#; b#; c#
a# = v# - r#
b# = store# - r#
c# = v# - store#
IF v# - r# >= 99 THEN store# = v#
REM IF v# - r# >= 99 THEN r# = r# + INT(b# / 2) + INT(c# / 2)
IF v# - r# >= 99 THEN r# = r# + 100
REM IF v# - r# >= 99 THEN r# = r# + (b# * 2)
```

```
REM IF v# - r# >= 99 THEN r# = r# + c#

a$ = INKEY$
IF a$ = "s" THEN STOP
LOOP
```

"C:\Users\Reactor1967\vcns\work\032403A.BAS"

```
v# = 1
r# = 1
store# = 1
CLS
RANDOMIZE TIMER
avg# = 1
count = 1
DO
z = INT(RND * 2) + 0
IF z = 0 THEN v# = v# + (v# - r# + 1)
IF z = 1 THEN v# = v# + (v# - r# + 2)
REM PRINT , , z; v#; r#; v# - r#; v# - store#; INT(avg# / count)
PRINT , , z; v#; r#; INT(avg# / count); INT(avg# / count) - x1#
x1# = INT(avg# / count)
a$ = INKEY$
```

```
REM INPUT a$

IF a$ = "s" THEN STOP

IF v# - store# >= 50 THEN flag = 1 ELSE flag = 0

IF flag = 1 THEN r# = r# + (v# - store#)

IF flag = 1 THEN store# = v#

IF flag = 1 THEN avg# = avg# + store#

IF flag = 1 THEN count = count + 1

LOOP
```

```
"C:\Users\Reactor1967\vcns\work\032303A.BAS"

v# = 1

r# = 1

store# = 1

z = 0

CLS

RANDOMIZE TIMER

DO

IF flag = 1 THEN z = 1 ELSE z = 0

IF z = 0 THEN v# = v# + (v# - r# + 1)

IF z = 1 THEN v# = v# + (v# - r# + 2)

PRINT , , z; v#; r#; v# - r#; v# - store#; store# - r#

a$ = INKEY$
```

```
IF a$ = "s" THEN STOP

IF a$ = "d" THEN EXIT DO

IF (v# - store#) + (store# - r#) >= 99 THEN flag = 1 ELSE flag = 0

IF flag = 1 THEN store# = v#

IF flag = 1 THEN r# = r# + 100

LOOP

DO

z = (v# / 2) - INT(v# / 2)

IF z = .5 THEN z = 1

PRINT , , z; v#; r#; v# - r#; (v# - store#) + (store# - r#); store#

test = (v# = 1) AND (r# = 1)

IF test = -1 THEN STOP

a$ = INKEY$

REM INPUT a$

IF a$ = "s" THEN STOP

dist# = (v# - r#)

IF (dist# / 2) - INT(dist# / 2) = .5 THEN dist# = dist# - 1

dist# = dist# / 2

dist# = dist# + 1

v# = v# - dist#

IF z = 1 THEN r# = r# - 100

IF z = 1 THEN store# = v#

LOOP
```

```
"C:\Users\Reactor1967\vcns\work\032203B.BAS"
v# = 1
r# = 1
store# = v#
CLS
RANDOMIZE TIMER
DO
z = INT(RND * 2) + 0
IF z = 0 THEN v# = v# + (v# - r# + 1)
IF z = 1 THEN v# = v# + (v# - r# + 2)
PRINT , , z; v#; r#; v# - store#; v# - r#; store# - r#; flag
REM INPUT a$
IF a$ = "s" THEN STOP
ts = (a$ = "d") AND (flag = 1)
IF ts = -1 THEN EXIT DO
REM IF v# - r# >= 99 THEN store# = v#
REM IF v# - r# >= 99 THEN r# = r# + 100
IF (v# - store#) + (store# - r#) >= 99 THEN flag = 1 ELSE flag = 0
IF flag = 1 THEN r# = r# + 100
IF flag = 1 THEN store# = v#
a$ = INKEY$
LOOP
```

```
store# = v#

REM r# = r# - 100

PRINT
"=====================================================================
="

DO

test# = (v# / 2) - INT(v# / 2)

IF test# = 0 THEN z = 0 ELSE z = 1

PRINT , , z; v#; r#; store# - v#; v# - r#; store# - r#; flag

INPUT a$

IF a$ = "s" THEN STOP

REM -------------decoding v#------------------------------

dist# = (v# - r#)

IF (dist# / 2) - INT(dist# / 2) = .5 THEN dist# = dist# - 1

dist# = dist# / 2

dist# = dist# + 1

v# = v# - dist#

REM -------------------------------------------------------

IF ((store# - v#) * 2) - (z - 1) >= 299 THEN flag = 1 ELSE flag = 0

IF flag = 1 THEN r# = r# - 100

LOOP
```

"C:\Users\Reactor1967\vcns\work\032203A.BAS"

```
v# = 1

r# = 1

store# = v#

CLS

RANDOMIZE TIMER

DO

z = INT(RND * 2) + 0

IF z = 0 THEN v# = v# + (v# - r# + 1)

IF z = 1 THEN v# = v# + (v# - r# + 2)

PRINT , , z; v#; r#; v# - store#; v# - r#; store# - r#; flag

REM IF v# - r# >= 99 THEN store# = v#

REM IF v# - r# >= 99 THEN r# = r# + 100

IF (v# - store#) + (store# - r#) >= 99 THEN flag = 1 ELSE flag = 0

IF flag = 1 THEN r# = r# + 100

IF flag = 1 THEN store# = v#

a$ = INKEY$

INPUT a$

IF a$ = "s" THEN STOP

LOOP
```

```
"C:\Users\Reactor1967\vcns\work\031603D.BAS"

v# = 99
```

```
r# = 1

store# = v#

CLS

DO

z = INT(RND * 2) + 0

test = ((v# - store#) >= 49) AND (r# + 100) <= v#

IF test = -1 THEN flag = 1 ELSE flag = 0

IF flag = 1 THEN r# = r# + 101

IF flag = 1 THEN store# = v#

IF z = 0 THEN v# = v# + (v# - r# + 1)

IF z = 1 THEN v# = v# + (v# - r# + 2)

PRINT , , z; v#; r#; v# - r#

REM IF r# > v# THEN STOP

REM IF v# - r# >= 200 THEN STOP

a$ = INKEY$

REM INPUT a$

IF a$ = "s" THEN STOP

LOOP
```

```
"C:\Users\Reactor1967\vcns\work\031603C.BAS"

REM Note: when then last v# of a specific r# minus then first v# of the

REM same r# is > 50 then r# increments.
```

```
REM Vn sub r minus Vn sub r => 50 then r = r + I

avg# = 0

avg1# = 0

avg2# = 0

v# = 1

r# = 1

store# = v#

CLS

DO

z = INT(RND * 2) + 0

v# = v# + (v# - r#) + z + 1

IF r# <> rs# THEN flag = 1 ELSE flag = 0

IF flag = 1 THEN avg# = avg# + (v# - store#)

IF flag = 1 THEN avg1# = avg1# + 1

IF flag = 1 THEN avg2# = avg2# + (v# - r#)

PRINT , INT(avg2# / avg1#); z; v#; r#; v# - r#; v# - store#; flag;
INT(avg# / avg1#)

rs# = r#

a$ = INKEY$

REM INPUT a$

IF a$ = "s" THEN STOP

IF (v# - r#) >= 99 THEN store# = v#

IF (v# - r#) >= 99 THEN r# = r# + 100

REM ---------------------------------------------
```

```
REM IF v# - store# >= 100 THEN flag = 1 ELSE flag = 0
REM IF flag = 1 THEN r# = r# + 100
REM IF flag = 1 THEN store# = v#
LOOP
```

"C:\Users\Reactor1967\vcns\work\031603B.BAS"

```
REM Note: when then last v# of a specific r# minus then first v# of the
REM same r# is > 50 then r# increments.
REM Vn sub r minus Vn sub r => 50 then r = r + I
avg# = 0
avg1# = 0
avg2# = 0
v# = 1
r# = 1
store# = v#
CLS
DO
z = INT(RND * 2) + 0
v# = v# + (v# - r#) + z + 1
IF r# <> rs# THEN flag = 1 ELSE flag = 0
IF flag = 1 THEN avg# = avg# + (v# - store#)
IF flag = 1 THEN avg1# = avg1# + 1
```

```
IF flag = 1 THEN avg2# = avg2# + (v# - r#)

PRINT , INT(avg2# / avg1#); z; v#; r#; v# - r#; v# - store#; flag;
INT(avg# / avg1#)

rs# = r#

a$ = INKEY$

REM INPUT a$

IF a$ = "s" THEN STOP

IF (v# - r#) >= 99 THEN store# = v#

IF (v# - r#) >= 99 THEN r# = r# + 100

REM ------------------------------------------------

REM IF v# - store# >= 100 THEN flag = 1 ELSE flag = 0

REM IF flag = 1 THEN r# = r# + 100

REM IF flag = 1 THEN store# = v#

LOOP
```

```
"C:\Users\Reactor1967\vcns\work\031603A.BAS"

r# = 1

v# = 1

t# = 0

CLS

RANDOMIZE TIMER

DO
```

```
t# = t# + 1

z = INT(RND * 2) + 0

dist# = (v# - r#)

IF z = 0 THEN v# = v# + (v# - r# + 1)

IF z = 1 THEN v# = v# + (v# - r# + 2)

PRINT z; v#; r#; v# - r#; dist#; v# - sv#; r# - sr; INT(v# / t#) - INT(r# /
t#)

sv# = v#

sr# = r#

lb# = (v# / t#) - 1

r# = lb# * t#

a$ = INKEY$

REM INPUT a$

IF a$ = "s" THEN STOP

LOOP
```

```
"C:\Users\Reactor1967\vcns\work\031503A.BAS"

r# = r# + dist# + N1

a# = 100

b# = 100

v# = 0

CLS
```

```
RANDOMIZE TIMER
DO
z = INT(RND * 2) + 0
IF z = 0 THEN v# = v# + 100 + z + 1
IF z = 1 THEN v# = v# + 100 + z + 2
b# = b# + 100 + 1
PRINT v#; b#; v# - b#
a$ = INKEY$
IF a$ = "s" THEN STOP
LOOP
```

"C:\Users\Reactor1967\vcns\work\031103A.BAS"

```
REM here I am using the speed of v# in relation with the speed of r#
REM logic here is that if you always know the speed of your v# then you
REM know the speed of your r#. If you know the time and speed
r# = 1
v# = 1
CLS
RANDOMIZE TIMER
count = 0
```

```
DO

count = count + 1

z = INT(RND * 2) + 0

v# = v# + (v# - r#) + z + 1

x1# = INT(v# / count)

x2# = INT(r# / count)

PRINT z; v#; r#; v# - r#; INT(v# / count); INT(v# / count) - x3#; INT(r# / count); INT(r# / count) - x4#; count; x1# - x2#

x3# = x1#

x4# = x2#

a$ = INKEY$

REM INPUT a$

IF a$ = "s" THEN STOP

REM IF (v# - r#) >= 99 THEN r# = r# + 100

dist# = INT(v# / count)

dist# = dist# - 1

r# = INT(dist# * count)

IF (r# / 2) - INT(r# / 2) = 0 THEN r# = r# - 1

LOOP
```

"C:\Users\Reactor1967\vcns\work\031003A.BAS"

```
REM SPEED OF R TESTING. TRYING TO GET A PATTERN TO
INCREASE THE SPEED OF R

REM FOR CODING BETWEEN VARIOUS BINARY NUMBERS

v# = 4

r# = 1

CLS

RANDOMIZE TIMER

DO

REM z = INT(RND * 2) + 0

REM IF z = 0 THEN z = 1 ELSE z = 0

z = 0

REM z = 1

dist# = (v# - r#)

REM dist# = dist# + 2

REM do stuff here.

dist# = dist# - 1

dist# = dist# / 2

r# = v# - dist#

IF (r# / 2) - INT(r# / 2) = 0 THEN r# = r# - 1

v# = v# + (v# - r#) + z + 1

PRINT , , z; v#; r#; v# - r#; v# - sv#; r# - sr#

sv# = v#

sr# = r#

a$ = INKEY$
```

```
INPUT a$

IF a$ = "s" THEN STOP

LOOP
```

"C:\Users\Reactor1967\vcns\work\030903A.BAS"

```
REM
*************************************************************************

REM ----------------STUDY THIS IT MIGHT BE A CANADATE
------------------------

REM ++++++++++++++++++++++++++++++++++++++++++++++++++++
+++++++++++++++++++++++++

REM new equation for doing stuff with the speed of r#

REM STUDY WHAT YOU KNOW TO DETERMINE IF THE
DISTANCE V2# - V1# - N IS THE

REM CURRENT SPEED OF R OR IF YOU NEED TO ADD 1 TO GET
THE SPEED OF R#

REM THINGS TO STUDY OR n, EVEN OR ODD (V# - R#), EVEN OR
ODD (V2# - V1# - n)

REM SEE IF THIS STUFF CAN TELL YOU HOW TO FIGURE OUR
SPEED OF R#.

d# = 1

v# = 1

RANDOMIZE TIMER

CLS
```

```
DO

z = INT(RND * 2) + 0

dist# = (v# - r#) + 1

IF z = 1 THEN dist# = dist# + 1

IF z = 2 THEN dist# = dist# + 2

IF z = 0 THEN r# = v# - INT((dist# - 1) / 2)

IF z = 1 THEN r# = v# - INT((dist# - 2) / 2)

dist# = (v# - r#)

IF z = 0 THEN v# = v# + (v# - r#) + 1

IF z = 1 THEN v# = v# + (v# - r#) + 2

PRINT z; v#; r#; v# - r#; (v# - r#) - x2#; dist#; v# - sv#; r# - sr#; (r# - sr#)
- x1#; (r# - sr#) - dist#

x1# = (r# - sr#)

x2# = (v# - r#)

a$ = INKEY$

IF a$ = "s" THEN STOP

REM d# = d# + 1

sv# = v#

sr# = r#


LOOP
```

"C:\Users\Reactor1967\vcns\work\030703C.BAS"

```
v# = 1

dist# = 100

RANDOMIZE TIMER

CLS

DO

z = INT(RND * 2) + 0

IF (v# / 2) - INT(v# / 2) = .5 THEN dist# = 100

IF (v# / 2) - INT(v# / 2) = 0 THEN dist# = 99

r# = v# - dist#

IF z = 0 THEN v# = v# + dist# + 1

IF z = 1 THEN v# = v# + dist# + 2

PRINT z; v#; r#; v# - r#; v# - sv#; r# - sr#

a$ = INKEY$

IF a$ = "s" THEN STOP

sv# = v#

sr# = r#

LOOP
```

"C:\Users\Reactor1967\vcns\work\030703B.BAS"

```
REM this can be used in any base. It stands to reason that if you know
your
```

```
REM distance of (v1# - r1#) and (v2# - r2#) that you can use these two

REM measurements for calculate the speed of r# (r2# - r1#). So a method

REM needs to be constructed that allows you to control the distances
between

REM (v1# - r1#) and (v2# - r2#) so that you can always calculate your

REM return speed of r# (r2# - r1#) for decode.

v# = 1

R# = 1

CLS

RANDOMIZE TIMER

DO

s1# = (v# - R#)

z = INT(RND * 2) + 0

v# = v# + ((v# - R#) * 1) + z + 1

PRINT z; v#; R#; v# - R#; s1#; v# - sv#; R# - sr#

a$ = INKEY$

REM INPUT a$

sv# = v#

sr# = R#

IF a$ = "s" THEN STOP

dist# = (v# - R#) - s1#

IF (dist# / 2) - INT(dist# / 2) = .5 THEN dist# = dist# - 1

IF dist# > 0 THEN R# = R# + dist#

LOOP
```

```
"C:\Users\Reactor1967\vcns\work\030703A.BAS"

CLS

v# = 101

d# = 18

RANDOMIZE TIMER

DO

z = INT(RND * 2) + 0

r# = v# - d#

v# = v# + (v# - r#) + z + 1

PRINT z; v#; r#; v# - r#; (v# - r#) - z4#; d#; (d# - z3#); v# - sv#; (v# -
sv#) - z1#; r# - sr#; (r# - sr#) - z2#

z1# = (v# - sv#)

z2# = (r# - sr#)

z3# = d#

z4# = (v# - r#)

sv# = v#

sr# = r#

a$ = INKEY$

INPUT a$

IF a$ = "s" THEN STOP

d# = d# + 1
```

```
x1# = (v# - r#)

test1 = (x1# / 2) - INT(x1# / 2) = .5

test2 = (d# / 2) - INT(d# / 2) = .5

test3 = (test1 = 0) AND (test2 = 0)

test4 = (test1 = 0) AND (test2 = -1)

test5 = (test1 = -1) AND (test2 = 0)

test6 = (test1 = -1) AND (test2 = -1)

IF test4 = -1 THEN d# = d# + 1

IF test5 = -1 THEN d# = d# + 1

LOOP
```

```
"C:\Users\Reactor1967\vcns\work\030603A.BAS"

v# = 10

r# = 1

CLS

RANDOMIZE TIMER

DO

z = INT(RND * 2) + 0

v# = v# + (v# - r#) + z + 1

PRINT z; v#; r#; v# - r#; v# - sv#; (v# - sv#) - x2#; r# - sr#; (r# - sr#) -
x1#

x1# = (r# - sr#)
```

```
x2# = (v# - sv#)

sv# = v#

sr# = r#

a$ = INKEY$

REM INPUT a$

IF a$ = "s" THEN STOP

a# = v# - r#

a# = (a# / 4)

a# = INT(a#)

a# = (a# * 2)

r# = r# + a#

LOOP
```

"C:\Users\Reactor1967\vcns\work\030503C.BAS"

```
REM unfinished

v# = 100

d1# = 1

d2# = 2

a# = (d1# * 2) + 1

b# = (d1# * 2) + 2

c# = (d2# * 2) + 1

d# = (d2# * 2) + 2
```

```
CLS

RANDOMIZE TIMER

DO

z = INT(RND * 2) + 0

test1# = (v# - d1#)

test2# = (v# - d2#)

test3# = (test1# / 2) - INT(test1# / 2)

test4# = (test2# / 3) - INT(test2# / 2)

test5 = (test3# = .5)

test6 = (test4# = .5)

IF test5 = -1 THEN dist# = d1#

IF test6 = -1 THEN dist# = d2#

r# = v# - dist#

v# = v# + (v# - r#) + z + 1

PRINT z; v#; r#; v# - r#; "|"; d1#; d2#; a#; b#; c#; d#

a$ = INKEY$

IF a$ = "s" THEN STOP

d1# = d1# + 2

d2# = d2# + 2

a# = (d1# * 2) + 1

b# = (d1# * 2) + 2

c# = (d2# * 2) + 1

d# = (d2# * 2) + 2
```

```
LOOP
```

```
"C:\Users\Reactor1967\vcns\work\030503B.BAS"
v# = 100
a# = 5
b# = 6
CLS
RANDOMIZE TIMER
DO
z = INT(RND * 2) + 0
IF z = 0 THEN dist# = (a# - 1) / 2
IF z = 1 THEN dist# = (b# - 2) / 2
r# = v# - dist#
v# = v# + (v# - r#) + z + 1
cal# = (a# - 1) / 2
sr1# = a2# - cal#
sr2# = b2# - cal#
PRINT z; v#; r#; "|"; a#; b#; sr1#; sr2#; "|"; v# - r#; r# - sr#
sv# = v#
sr# = r#
a$ = INKEY$
IF a$ = "s" THEN STOP
```

```
a2# = a#

b2# = b#

a# = a# + 2

b# = b# + 2

LOOP
```

"C:\Users\Reactor1967\vcns\work\030503A.BAS"

```
REM the lower part of this program is worth study. This can almost be

REM predicted.

v# = 1

r# = 1

CLS

RANDOMIZE TIMER

DO

z = INT(RND * 2) + 0

dist# = (v# - r#)

IF z = 0 THEN v# = v# + (v# - r# + 1)

IF z = 1 THEN v# = v# + (v# - r# + 2)

PRINT , , z; v#; r#; v# - r#; dist#; v# - sv#; r# - sr#

sv# = v#

sr# = r#

a$ = INKEY$
```

```
REM INPUT a$
IF a$ = "s" THEN STOP
LOOP UNTIL (v# - r#) >= 1000
DO
z = INT(RND * 2) + 0
dist# = (v# - r#)
IF (dist# / 2) - INT(dist# / 2) = .5 THEN dist# = dist# - 1
dist# = dist# / 2
r# = (v# - dist#)
REM IF (r# / 2) - INT(r# / 2) = 0 THEN r# = r# + 1
IF z = 0 THEN v# = v# + (v# - r# + 1)
IF z = 1 THEN v# = v# + (v# - r# + 2)
PRINT z; v#; r#; v# - r#; dist#; v# - sv#; r# - sr#
sv# = v#
sr# = r#
a$ = INKEY$
REM INPUT a$
IF a$ = "s" THEN STOP
LOOP
```

```
"C:\Users\Reactor1967\vcns\work\030403B.BAS"

v# = 1

r# = 1

CLS

RANDOMIZE TIMER

DO

z = INT(RND * 2) + 0

IF flag = 1 THEN z = 3

a# = r# + z + 1

a# = (a# / 3) - INT(a# / 3)

a# = a# * 10

a# = INT(a#)

v# = v# + (v# - r#) + z + 1

sc# = v#

DO

b# = (v# / 3) - INT(v# / 3)

b# = b# * 10

b# = INT(b#)

IF b# = a# THEN EXIT DO

v# = v# + 1

a$ = INKEY$

IF a$ = "s" THEN STOP
```

```
REM PRINT a#; b#
LOOP
sc# = (v# - sc#)
PRINT , , z; v#; r#; v# - r#; sc#; v# - sv#; r# - sr#
sv# = v#
sr# = r#
a$ = INKEY$
REM INPUT a$
IF a$ = "s" THEN STOP
IF flag = 1 THEN r# = r# + 207
IF v# - r# >= 100 THEN flag = 1 ELSE flag = 0
LOOP
```

```
"C:\Users\Reactor1967\vcns\work\030403A.BAS"
v# = 1
r# = 1
CLS
DO
z = INT(RND * 3) + 0
a# = v# + ((v# - r#) * 2) + z + 1
a# = (a# / 3) - INT(a# / 3)
v# = v# + (v# - r#) + z + 1
```

```
sc# = v#

DO

b# = (v# / 3) - INT(v# / 3)

IF b# = a# THEN EXIT DO

v# = v# + 1

a$ = INKEY$

IF a$ = "s" THEN STOP

LOOP

sc# = (v# - sc#)

PRINT , , z; v#; r#; v# - r#; v# - sv#; r# - sr#; sc#

sv# = v#

sr# = r#

a$ = INKEY$

REM INPUT a$

IF a$ = "s" THEN STOP

IF v# - r# >= 100 THEN r# = r# + (v# - r#)

LOOP
```

```
"C:\Users\Reactor1967\vcns\work\030203C.BAS"

REM there seems to be a mathmatical progression of what you add to

REM v1# to get v2# when dividing v# / x for specific remainders to tell

REM what N is. You divide the distance by three to test it.
```

```
REM try using this in a base two system so can find way to test when

REM to increment and decrement r#.

REM v# = v# + (v# - r#) + N + (C1 or C2 or C3 or C4) to use other bases
remainders in other bases

REM c# = (correctional value for 0 to a 1 or 1 to a 0 or 1 to a 2 or 2 to a
1)

v# = 1

r# = 1

CLS

RANDOMIZE TIMER

DO

z = INT(RND * 3) + 0

IF z = 0 THEN a# = ((v# - r#) * 2) + 1

IF z = 1 THEN a# = ((v# - r#) * 2) + 2

IF z = 2 THEN a# = ((v# - r#) * 2) + 3

IF z = 0 THEN v# = v# + ((v# - r#) * 2) + 1

IF z = 1 THEN v# = v# + ((v# - r#) * 2) + 2

IF z = 2 THEN v# = v# + ((v# - r#) * 2) + 3

PRINT z; v#; r#; a#; (a# / 3) - INT(a# / 3)

a$ = INKEY$

INPUT a$

IF a$ = "s" THEN STOP

LOOP
```

```
"C:\Users\Reactor1967\vcns\work\030203B.BAS"

v# = 1

r# = 1

RANDOMIZE TIMER

CLS

DO

z = INT(RND * 2) + 0

IF z = 0 THEN v# = v# + (v# - r# + 1)

IF z = 1 THEN v# = v# + (v# - r# + 2)

PRINT , , (v# / 3) - INT(v# / 3); z; v#; r#; v# - r#

a$ = INKEY$

IF a$ = "s" THEN STOP

IF v# - r# >= 50 THEN STOP

LOOP
```

```
"C:\Users\Reactor1967\vcns\work\030203A.BAS"

REM in base to the equation is v# = v# + ((v# - r#) * 2) + N

REM ideal using base two increment r# so that a binary 0 / 3 has the

REM same remainder every time and a binary 1 / 3 has the same
remainder

REM every time so that when r# encreases or decreaces when use another
```

```
REM equation for base 3 to increment v# = v# ((v# - r#) * 2) + 3 so that

REM its remainder is different. When we decode we look for that different

REM remainder so we will know when to decrement r# and use the second decode

REM equation for the second encode equation.

REM This program here uses base 3 and does not apply to the ideal above it

REM just was used to gain the ideal above.

v# = 1

r# = 1

CLS

RANDOMIZE TIMER

DO

z = INT(RND * 3) + 0

dist# = v# - r#

v# = v# + ((v# - r#) * 2) + z + 1

REM PRINT v#; r#; v# - r#; dist#; v# - sv#; r# - sr#

PRINT z; v#; r#; v# - r#; v# / 3 - INT(v# / 3); (v# - r#) / 3

sv# = v#

sr# = r#

a$ = INKEY$

INPUT a$

IF a$ = "s" THEN STOP

test1 = ((v# - r#) >= 102) AND ((v# - r#) < 202)
```

```
test2 = ((v# - r#) >= 202) AND ((v# - r#) < 302)

test3 = ((v# - r#) >= 302) AND ((v# - r#) < 402)

IF test1 = -1 THEN r# = r# + 102

IF test2 = -1 THEN r# = r# + 202

IF test3 = -1 THEN r# = r# + 302

LOOP
```

```
"C:\Users\Reactor1967\vcns\work\030103A.BAS"

v# = 1

r# = 1

RANDOMIZE TIMER

CLS

DO

z = INT(RND * 3) + 0

gu# = (v# - r#)

IF z = 0 THEN v# = v# + ((v# - r#) * 2) + 1

IF z = 1 THEN v# = v# + ((v# - r#) * 2) + 2

IF z = 2 THEN v# = v# + ((v# - r#) * 2) + 3

PRINT , z; v#; r#; v# - r#; gu#; r# - sr#

sr# = r#

a$ = INKEY$

REM INPUT a$
```

```
IF a$ = "s" THEN STOP
dist# = v# - r#
dist# = INT(dist# / 3)
r# = v# - dist#
LOOP
```

"C:\Users\Reactor1967\vcns\work\022803C.BAS"

```
v# = 1
r# = 1
CLS
DO
a# = v# + ((v# - r#) * 2) + 1
b# = v# + ((v# - r#) * 2) + 2
c# = v# + ((v# - r#) * 2) + 3
PRINT r#, v#; a#; b#; c#
a$ = INKEY$
IF a$ = "s" THEN STOP
v# = v# + 1
IF (v# - r#) >= 102 THEN r# = r# + 102
LOOP
```

```
"C:\Users\Reactor1967\vcns\work\022803B.BAS"

a# = 1

b# = 2

c# = 3

d# = 4

count = 0

CLS

DO

PRINT (a# / 3); count; a#; b#; c#; d#

count = count + 1

IF count = 4 THEN count = 1

a# = a# + 1

b# = b# + 3

c# = c# + 3

d# = d# + 3

a$ = INKEY$

IF a$ = "s" THEN STOP

LOOP
```

```
"C:\Users\Reactor1967\vcns\work\022803A.BAS"

a# = 1
```

```
b# = 2

c# = 3

d# = 4

count = 0

CLS

DO

PRINT (a# / 3); count; a#; b#; c#; d#

count = count + 1

IF count = 4 THEN count = 1

a# = a# + 1

b# = b# + 3

c# = c# + 3

d# = d# + 3

a$ = INKEY$

IF a$ = "s" THEN STOP

LOOP
```

"C:\Users\Reactor1967\vcns\work\022603D.BAS"

```
r# = 1

v# = 1

range# = 102

CLS
```

```
RANDOMIZE TIMER

DO

z = INT(RND * 3) + 0

gh# = (v# - r#)

IF z = 0 THEN v# = v# + ((v# - r#) * 2) + 1

IF z = 1 THEN v# = v# + ((v# - r#) * 2) + 2

IF z = 2 THEN v# = v# + ((v# - r#) * 2) + 3

PRINT z; v#; r#; v# - r#; gh#; v# - sv#; r# - sr#; (r# - sr#) - x1#; range#

x1# = (r# - sr#)

sv# = v#

sr# = r#

IF v# - r# > range# THEN GOSUB align:

a$ = INKEY$

REM INPUT a$

IF a$ = "s" THEN STOP

LOOP

align:

range# = range# + 102

dist# = (v# - r#)

dist# = INT(dist# / 3)

dist# = dist# * 3

dist# = dist# - range#

r# = r# + dist#

RETURN
```

```
"C:\Users\Reactor1967\vcns\work\022603C.BAS"

r# = 1

a# = 1

b# = 2

c# = 3

d# = 4

range# = 100

RANDOMIZE TIMER

CLS

DO

z = INT(RND * 3) + 0

IF z = 0 THEN a# = a# + (a# - r#) + 1

IF z = 1 THEN a# = a# + (a# - r#) + 2

IF z = 2 THEN a# = a# + (a# - r#) + 3

PRINT z; a#; r#; a# - r#; range#

a$ = INKEY$

IF a$ = "s" THEN STOP

REM IF a$ = "g" THEN GOSUB spr:

IF a# - r# > range# THEN GOSUB spr:

LOOP

spr:
```

```
DO
r# = r# + 3
IF a# - r# < range# THEN EXIT DO
a$ = INKEY$
IF a$ = "s" THEN STOP
LOOP
range# = range# + 100
RETURN
```

```
"C:\Users\Reactor1967\vcns\work\022603B.BAS"
REM r# alignment program.
REM Conclusions. R# needs to be divisable by the base.
REM So, practice with two r#'s here.
REM 1. a control r#
REM 2. a experimental r#
REM see if can find a pattern to control the speed of r# in relation to
REM v# so that when decoding the speed of r# can be decoding.
REM it seems you want r# and v# to start far enough apart so that
REM r can change everytime v changes. So that a 0 to a 0 = change1 for r
REM ,0 to a 1 = change2 for r, 1 to a 0 = change3 for r, 1 to a 1 = change4
REM for 4. When decoding observe the N values of v and make the
appropiate
```

```
REM changes for the speed of r.
rc# = 1
a# = 1
b# = 2
c# = 3
d# = 4
expr# = 4
CLS
flag = 0
DO
REM PRINT flag; a#; b#; c#; d#; (b# / 3) - INT(b# / 3); (c# / 3) - INT(c# / 3); (d# / 3) - INT(d# / 3)
PRINT flag; expr#; rc#; a#; b#; c#; d#; a# - rc#
IF a# - rc# >= expr# THEN lb = 1 ELSE lb = 0
flag = flag + 1
IF flag = 4 THEN flag = 1
IF lb = 1 THEN expr# = expr# + 3
IF lb = 1 THEN rc# = rc# + 3
a# = a# + 1
b# = b# + 3
c# = c# + 3
d# = d# + 3
a$ = INKEY$
REM INPUT a$
```

```
IF a$ = "s" THEN STOP
LOOP
```

"C:\Users\Reactor1967\vcns\work\022603a.txt"

1 = 2 3 4

2 = 5 6 7

3 = 8 9 10

4 = 11 12 13

5 = 14 15 16

7 = 17 18 19

What you need is to increase r# by the base till its with in the range of v# that you want it. So practice taking r# with in range of each value of n to determine what the speed of r# will be when coding between different values of N and what the speed of r# will be when decoding specific values of N. When you decode deteremine the current value of N and the previous value of N then look at the decode chart for the speed of r# to deteremine what you subtract from the speed of r# to get the previous value of the speed of r#. In fact, you have to keep control up what the speed of r# is at all times. while coding and decoding.

```
"C:\Users\Reactor1967\vcns\work\022603A.BAS"

v# = 1

r# = 1

CLS

RANDOMIZE TIMER

DO

z = INT(RND * 2) + 0

IF z = 0 THEN v# = v# + (v# - r# + 4)

IF z = 1 THEN v# = v# + (v# - r# + 8)

IF z = 0 THEN gy# = gy# + 4

IF z = 1 THEN gy# = gy# + 8

PRINT z; v#; r#; v# - r#; v# - sv#; r# - sr#; (r# - sr#) - x1#

REM PRINT z; v# / 8; (r# + 4) / 8; (r# + 8) / 8

REM -----------------

REM testing variables

x1# = (r# - sr#)

sv# = v#

sr# = r#

REM -----------------

REM ----------------------------------

REM testing r# for alignment. Programe will abort if out of alignment

a# = (r# + 4) / 8: b# = (r# + 8) / 8

a# = a# - INT(a#): b# = b# - INT(b#)
```

```
n1t# = (v# / 8) - INT(v# / 8)

n2t = (z = 0) AND (n1t# = a#)

n3t = (z = 1) AND (n1t# = b#)

n4t = (n2t = -1) OR (n3t = -1)

IF n4t = 0 THEN STOP

REM ---------------------------------

REM program halt sequence

a$ = INKEY$

IF a$ = "d" THEN EXIT DO

IF a$ = "s" THEN STOP

REM ------------------------------

REM - r# control sequence here.

x# = (v# - r#) / 2

r# = v# - x#

IF z = 0 THEN r# = r# - 2

REM ------------------------------


LOOP
```

"C:\Users\Reactor1967\vcns\work\022503B.BAS"

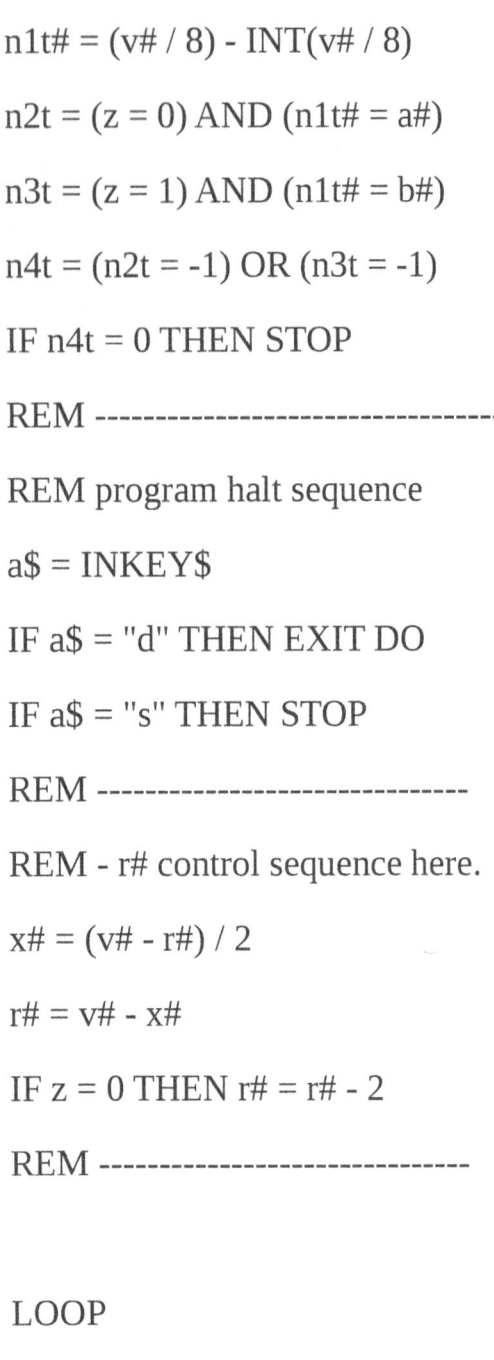

```
REM ----------DECODE KEY FOR DECREMENTING THE SPEED OF
R#------------------
```

```
REM speed of r# in this program is the (r# - sr#) value
REM binary 0 to a binary 0 = 2
REM binary 0 to a binary 1 = 4
REM binary 1 to a binary 0 = 2
REM binary 1 to a binary 1 = 4
REM -----------------------------------------------------------
v# = 1
r# = 1
CLS
RANDOMIZE TIMER
DO
z = INT(RND * 2) + 0
IF z = 0 THEN v# = v# + (v# - r# + 4)
IF z = 1 THEN v# = v# + (v# - r# + 8)
IF z = 0 THEN gy# = gy# + 4
IF z = 1 THEN gy# = gy# + 8
PRINT z; v#; r#; v# - r#; (v# - r#) - x2#; v# - sv#; r# - sr#; (r# - sr#) - x1#
REM PRINT z; v# / 8; (r# + 4) / 8; (r# + 8) / 8
a$ = INKEY$
IF a$ = "d" THEN EXIT DO
IF a$ = "s" THEN STOP
x1# = (r# - sr#)
x2# = (v# - r#)
sv# = v#
```

```
sr# = r#

dist# = (v# - r#)

dist# = dist# / 2

r# = v# - dist#

LOOP
```

"C:\Users\Reactor1967\vcns\work\022503A.BAS"

```
REM this program could decode but like my other programs has a if

REM if you can figure out the N value. Do this by dividing v# into 8

REM add n1 to r# and divid by 8 then add n2 to r# and divide by 8

REM ignore the last two values and look at the last 3rd value. It

REM should tell you if you have a binary 1 or a binary 0.

REM You use the predictable rate of change in the speed of r# as a tool

REM to decode.

REM ------------------------------------------------------------

REM ----------DECODE KEY FOR DECREMENTING THE SPEED OF R#-----------------

REM rate of change better known as a derivative for speed of r#

REM speed of r# in this program is the (r# - sr#) value

REM binary 0 to a binary 0 = 2

REM binary 0 to a binary 1 = 4

REM binary 1 to a binary 0 = 2
```

```
REM binary 1 to a binary 1 = 4

REM -------------------------------------------------------------

REM what you need to work on is knowing the predictable rate of change for

REM the speed of r and finding your n value.

REM secret here is that your distance for (v2# - v1#) always has to

REM increase at the same rates for there specific n values and

REM distance for (r2# - r1#) always has to increase at the same

REM rates for there specific N values in relation to the previous (v2# - r2#)

REM because the previous (v2# - r2#) - (v2# - v1#) - N2 = (r2# - r1#)

v# = 1

r# = 1

CLS

RANDOMIZE TIMER

DO

z = INT(RND * 2) + 0

REM v# = v# + (v# - r#) + z + 1

gnk# = (v# - r#)

IF z = 0 THEN v# = v# + (v# - r# + 4)

IF z = 1 THEN v# = v# + (v# - r# + 8)

REM PRINT z; v#; r#; gnk#; gnk# - lb#; (v# - r#); (v# - r#) - x#; v# - sv#; (v# - sv#) - tv#; r# - sr#; (r# - sr#) - tr#; (v# - sv#) + (r# - sr#); (v# - sv#) + (r# - sr#) - gly#

REM PRINT z; v#; r#; gnk#; (v# - r#); v# - sv#; r# - sr#
```

```
REM PRINT z; v#; r#; gnk#

PRINT z; v#; r#; v# - r#; v# - sv#; r# - sr#

lb# = gnk#

tv# = (v# - sv#)

tr# = (r# - sr#)

gly# = (v# - sv#) + (r# - sr#)

x# = v# - r#

sv# = v#

sr# = r#

a$ = INKEY$

IF a$ = "s" THEN STOP

dist# = (v# - r#)

IF (dist# / 2) - INT(dist# / 2) = .5 THEN flag = 1 ELSE flag = 0

IF flag = 1 THEN dist# = dist# - 1

dist# = dist# / 2

IF flag = 1 THEN dist# = dist# + 1

r# = v# - dist#

REM IF (r# / 2) - INT(r# / 2) = 0 THEN r# = r# - 1

LOOP
```

"C:\Users\Reactor1967\vcns\work\022303E.BAS"

```
RANDOMIZE TIMER
```

```
CLS
redo:
v# = 1
r# = 1
DO
z = INT(RND * 2) + 0
v# = v# + (v# - r#) + z + 1
a$ = INKEY$
IF a$ = "s" THEN STOP
LOOP UNTIL v# - r# >= 100
REM PRINT , v#; r#; v# - r#
a$ = INKEY$
IF a$ = "s" THEN STOP
dist# = v# - r#
sv# = 0
sr# = 0
d# = 0
PRINT z; dist#; d#; sv#; sr#, v#; r#
DO
z = INT(RND * 2) + 0
REM -----------------------------
IF z = 0 THEN d# = d# + 1
IF z = 1 THEN d# = d# + 2
sr# = dist# - d#
```

```
sv# = d#
REM ----------------------------
REM sr# = sr# + 2
REM sv# = dist# - sr#
REM d# = dist# - sr#
REM ----------------------------
IF z = 0 THEN sv# = sv# + 1
IF z = 1 THEN sv# = sv# + 2
dist# = (d# * 2)
IF z = 0 THEN dist# = dist# + 1
IF z = 1 THEN dist# = dist# + 2
v# = v# + sv#: r# = r# + sr#
PRINT z; dist#; d#; sv#; sr#; (sr# - x1#); v#; r#
x1# = sr#
a$ = INKEY$
REM INPUT a$
IF a$ = "s" THEN STOP
LOOP
```

"C:\Users\Reactor1967\vcns\work\022303D.BAS"

```
r# = 2
v# = 2
```

```
RANDOMIZE TIMER

CLS

DO

z = INT(RND * 2) + 0

test = (v# + (v# - r# + 6)) - r# >= 100

IF test = -1 THEN z = 3

IF z = 0 THEN v# = v# + (v# - r# + 2)

IF z = 1 THEN v# = v# + (v# - r# + 4)

IF z = 3 THEN v# = v# + (v# - r# + 6)

PRINT , , z; v#; r#; v# - r#

REM IF (v# - r#) < 0 THEN STOP

a$ = INKEY$

REM INPUT a$

IF a$ = "s" THEN STOP

IF test = -1 THEN r# = r# + 104

LOOP
```

```
"C:\Users\Reactor1967\vcns\work\022303C.BAS"

DECLARE SUB test (v#, r#, z!)

v# = 10000

r# = 9000

CLS
```

```
RANDOMIZE TIMER

DO

z = INT(RND * 2) + 0

CALL test(v#, r#, z)

IF z = 0 THEN v# = v# + (v# - r# + 2)

IF z = 1 THEN v# = v# + (v# - r# + 4)

PRINT , , v#; r#; v# - r#; v# - sv#; r# - sr#

sv# = v#

sr# = r#

a$ = INKEY$

IF a$ = "s" THEN STOP

LOOP


SUB test (v#, r#, z)

dist# = v# - r#

rt# = r#

DO

rt# = rt# + 2

test3# = (v# - r#)

IF z = 0 THEN test3# = test3# + 2

IF z = 1 THEN test3# = test3# + 4

test2 = (v# + test3#) - rt# = ((v# + test3#) - v#) + (rt# - r#)

IF test2 = -1 THEN EXIT DO

a$ = INKEY$
```

```
IF a$ = "s" THEN STOP

IF rt# > v# THEN EXIT DO

LOOP

IF test2 = -1 THEN r# = rt#

END SUB
```

"C:\Users\Reactor1967\vcns\work\022303B.BAS"

```
DECLARE SUB test (v#, r#, z!)

REM (v2 - r2) must equal = (v2 - v1) + (r2 - r1)

REM rate of change of (v2 - r2) must equal = rate of change of (v2 - v1)
+ (r2 - r1)

REM Conclusion it seems maybe I should not stick with specific values
for n

REM but rather different values for N and modulate the distance between

REM v# and r# much like radio signals are modulated so that by reading

REM the distances back between v# and r# data can be incoded and
decoded

REM and instructions for incrementing/decrement r# can be incoded and

REM decoded as well. Humm. Much needed thought here.

CLS

RANDOMIZE TIMER

v# = 10000

r# = 8496
```

```
DO

z = INT(RND * 2) + 0

ghu# = r#

CALL test(v#, r#, z)

gk# = (v# - r#)

v# = v# + (v# - r#) + z + 0

PRINT , z; v#; r#; v# - r#; gk#; v# - sv#; r# - sr#

sv# = v#

IF r# - ghu# > 0 THEN sr# = r#

a$ = INKEY$

IF a$ = "s" THEN STOP

LOOP


SUB test (v#, r#, z)

dist# = (v# - r#)

dist2# = (v# - r#)

d2# = 1

IF (dist# / 2) - INT(dist# / 2) = .5 THEN dist# = dist# - 1

dist# = dist# / 2

dist# = dist# + 100

FOR count = 1 TO dist2#

d# = dist# - d2#

sr# = dist# - d#

sv# = dist# + z + 1
```

```
test2 = (v# + sv#) - (r# + sr#) = (sv# + sr#)

IF test2 = -1 THEN EXIT FOR

d2# = d2# + 1

a$ = INKEY$

IF a$ = "s" THEN STOP

NEXT

IF test2 = -1 THEN r# = v# - d#

REM IF test2 = 0 THEN r# = v# - (v# - r#)

END SUB
```

```
"C:\Users\Reactor1967\vcns\work\022303A.BAS"

d# = 2

RANDOMIZE TIMER

v# = 100

CLS

DO

d# = d# + 2

x# = d# / 2

IF (x# / 2) - INT(x# / 2) = .5 THEN d# = d# + 2

x# = d# / 2

r# = v# - x#

z = INT(RND * 2) + 0
```

```
IF z = 0 THEN v# = v# + (v# - r# + 2)

IF z = 1 THEN v# = v# + (v# - r# + 4)

PRINT z; v#; r#; v# - r#; (v# - r#) - x5#; v# - sv#; (v# - sv#) - x1#; r# - sr#; (r# - sr#) - x2#; (v# - sv#) + (r# - sr#); (v# - sv#) + (r# - sr#) - x3#

x1# = (v# - sv#)

x2# = (r# - sr#)

x3# = (v# - sv#) + (r# - sr#)

x5# = (v# - r#)

sv# = v#

sr# = r#

a$ = INKEY$

IF a$ = "s" THEN STOP

LOOP
```

"C:\Users\Reactor1967\vcns\work\022303.txt"

2

2 = 2 4 6

4 = 8 10 12

6 = 12 14 16

8 = 16 18 20

10 = 20 22 24

12 = 24 26 28

```
"C:\Users\Reactor1967\vcns\work\022203E.BAS"

v# = 2

r# = 1

d# = 0

spr# = 0

CLS

RANDOMIZE TIMER

DO

z = INT(RND * 2) + 0

v# = v# + (v# - r#) + z + 1

PRINT z; v#; r#; v# - r#; (v# - r#) - x4#; v# - sv#; (v# - sv#) - x1#; r# - sr#; (r# - sr#) - x2#; (v# - sv#) + (r# - sr#); (v# - sv#) + (r# - sr#) - x3#

x1# = (v# - sv#)

x2# = (r# - sr#)

x3# = (v# - sv#) + (r# - sr#)

x4# = (v# - r#)

sv# = v#

sr# = r#

a$ = INKEY$

REM INPUT a$

IF a$ = "s" THEN STOP

IF z = 0 THEN d# = d# + 1 ELSE d# = d# + 2
```

```
r# = v# - d#
REM -------------------------
REM IF z = 0 THEN spr# = spr# + 1
REM IF z = 1 THEN spr# = spr# + 2
REM r# = r# + spr#
LOOP
```

```
"C:\Users\Reactor1967\vcns\work\022203D.BAS"
REM this program could decode but like my other programs has a if
REM if you can figure out the N value.
REM You use the predictable rate of change in the speed of r# as a too
REM to decode.
REM -----------------------------------------------------------
REM rate of change better known as a derivative for speed of r#
REM speed of r# in this program is the (r# - sr#) value
REM binary 0 to a binary 0 = 1
REM binary 0 to a binary 1 = 1
REM binary 1 to a binary 0 = 2
REM binary 1 to a binary 1 = 2
REM -----------------------------------------------------------
REM what you need to work on is knowing the predictable rate of change
for
```

```
REM the speed of r and finding your n value.

REM secret here is that your distance for (v2# - v1#) always has to

REM increase at the same rates for there specific n values and

REM distance for (r2# - r1#) always has to increase at the same

REM rates for there specific N values in relation to the previous (v2# - r2#)

REM because the previous (v2# - r2#) - (v2# - v1#) - N2 = (r2# - r1#)

v# = 1

r# = 1

CLS

RANDOMIZE TIMER

DO

z = INT(RND * 2) + 0

REM v# = v# + (v# - r#) + z + 1

gnk# = (v# - r#)

IF z = 0 THEN v# = v# + (v# - r# + 2)

IF z = 1 THEN v# = v# + (v# - r# + 4)

PRINT z; v#; r#; gnk#; gnk# - lb#; (v# - r#); (v# - r#) - x#; v# - sv#; (v# - sv#) - tv#; r# - sr#; (r# - sr#) - tr#; (v# - sv#) + (r# - sr#); (v# - sv#) + (r# - sr#) - gly#

lb# = gnk#

tv# = (v# - sv#)

tr# = (r# - sr#)

gly# = (v# - sv#) + (r# - sr#)

x# = v# - r#
```

```
sv# = v#
sr# = r#
a$ = INKEY$
IF a$ = "s" THEN STOP
dist# = (v# - r#)
IF (dist# / 2) - INT(dist# / 2) = .5 THEN flag = 1 ELSE flag = 0
IF flag = 1 THEN dist# = dist# - 1
dist# = dist# / 2
IF flag = 1 THEN dist# = dist# + 1
r# = v# - dist#
REM IF (r# / 2) - INT(r# / 2) = 0 THEN r# = r# - 1
LOOP
```

```
"C:\Users\Reactor1967\vcns\work\022203C.BAS"
v# = 4
r# = 1
bin = 1
spr# = 0
CLS
RANDOMIZE TIMER
DO
z = INT(RND * 2) + 0
```

```
test1 = (z = 0) AND (bin = 0)

test2 = (z = 0) AND (bin = 1)

test3 = (z = 1) AND (bin = 0)

test4 = (z = 1) AND (bin = 1)

bin = z

IF test1 = -1 THEN spr# = spr# + 1

IF test2 = -1 THEN spr# = spr# + 1

IF test3 = -1 THEN spr# = spr# + 2

IF test4 = -1 THEN spr# = spr# + 2

v# = v# + (v# - r#) + z + n

PRINT , , v#; r#; v# - r#

a$ = INKEY$

INPUT a$

IF a$ = "s" THEN STOP

r# = r# + spr#

LOOP
```

"C:\Users\Reactor1967\vcns\work\022203B.BAS"

```
REM Here (v2# - r2#) - (v1# - r1#) when going from a binary 1 to a
binary 0

REM respectfully = 1

REM Here (v2# - r2#) - (v1# - r1#) when going from a binary 0 to a
binary 0
```

```
REM respectfully = 2

REM Here (v2# - r2#) - (v1# - r1#) when going from a binary 1 to a
binary 1

REM respectfully = 2

REM Here (v2# - r2#) - (v1# - r1#) when going from a binary 0 to a
binary 1

REM respectfully = 3

REM -------------------------------------------

REM rate of change for sv# and sr#

REM 0 to a 1 sv# = 2

REM 0 to a 1 sr# = 1

REM 0 to a 0 sv# = 1

REM 0 to a 0 sr# = 1

REM 1 to a 0 sv# = 0

REM 1 to a 0 sr# = 1 or 2

REM 1 to a 1 sv# = 1

REM 1 to a 1 sr# = 1 or 2

v# = 1

r# = 1

CLS

RANDOMIZE TIMER

DO

z = INT(RND * 2) + 0

v# = v# + (v# - r#) + z + 1
```

```
PRINT z; v#; r#; v# - r#; (v# - r#) - x#; v# - sv#; (v# - sv#) - tv#; r# - sr#;
(r# - sr#) - tr#; (v# - sv#) + (r# - sr#); (v# - sv#) + (r# - sr#) - gly#

tv# = (v# - sv#)

tr# = (r# - sr#)

gly# = (v# - sv#) + (r# - sr#)

x# = v# - r#

sv# = v#

sr# = r#

a$ = INKEY$

IF a$ = "s" THEN STOP

dist# = (v# - r#)

IF (dist# / 2) - INT(dist# / 2) = .5 THEN flag = 1 ELSE flag = 0

IF flag = 1 THEN dist# = dist# - 1

dist# = dist# / 2

IF flag = 1 THEN dist# = dist# + 1

r# = v# - dist#

LOOP
```

```
"C:\Users\Reactor1967\vcns\work\022203A.BAS"

dist# = 1650

spr# = dist#

CLS
```

```
RANDOMIZE TIMER
DO
x# = dist#
x2# = INT(dist# / 2) - 1
z = INT(RND * 2) + 0
DO
d# = x2#
r# = dist# - d#
spr# = d# + r# + z + 1
a$ = INKEY$
IF a$ = "s" THEN STOP
test = (spr# >= dist#) AND (d# > INT(dist# / 2))
IF test = -1 THEN EXIT DO
x2# = x2# + 1
LOOP
dist# = (d# * 2) + z + 1
PRINT , , z; d#; r#; dist#; dist# - g2#; spr#; spr# - g#
g# = spr#
g2# = dist#
a$ = INKEY$
INPUT a$
IF a$ = "s" THEN STOP
LOOP
```

"C:\Users\Reactor1967\vcns\work\022003G.BAS"

```
REM attempting to control a number that is the sum of (v2# - v1#) - (r2#
- r1#)

REM so that if you know any of the two the other one can be figured out.

V# = 1

R# = 1

d# = 0

CLS

RANDOMIZE TIMER

DO

z = INT(RND * 2) + 0

IF z = 0 THEN d# = d# + 1

IF z = 1 THEN d# = d# + 2

R# = V# - d#

V# = V# + (V# - R#) + z + 1

PRINT z; V#; R#; V# - R#; d#; V# - SV#; R# - SR#; (V# - SV#) + (R# -
SR#)

SV# = V#

SR# = R#

a$ = INKEY$

IF a$ = "s" THEN STOP

LOOP
```

```
"C:\Users\Reactor1967\vcns\work\022003F.BAS"
REM maybe it is possible to control the speed of v# instead the
REM speed of r#. Slow the speed of v# down to let r# catch up and
REM speed v# up when r# gets too close.
v# = 1
r# = 1
CLS
RANDOMIZE TIMER
DO
z = INT(RND * 2) + 0
gn# = v# - r#
v# = v# + (v# - r#) + z + 1
PRINT , , v#; r#; v# - r#; gn#; v# - sv#
sr# = r#
IF v# - sv# > 100 THEN GOSUB cr:
sv# = v#
a$ = INKEY$
IF a$ = "s" THEN STOP
LOOP
cr:
dist# = v# - r#
d# = 1
```

```
DO

x# = dist# - d#

d# = d# + 1

a$ = INKEY$

IF a$ = "s" THEN STOP

LOOP UNTIL x# < 25

r# = v# - x#

RETURN
```

"C:\Users\Reactor1967\vcns\work\022003E.BAS"

```
REM it seems when the speed of v# reaches a high end the speed of

REM r is increased. When decoding when the speed of v# reachs a low end

REM the speed of r is decreased.

v# = 1

r# = 1

CLS

RANDOMIZE TIMER

DO

z = INT(RND * 2) + 0

dist# = (v# - r#)

v# = v# + (v# - r#) + z + 1
```

```
PRINT , , z; v#; r#; v# - r#; dist#; r# - sr#
REM PRINT v# - sv#; r# - sr#
sv# = v#
sr# = r#
a$ = INKEY$
IF a$ = "s" THEN STOP
IF v# - r# >= 100 THEN r# = v# - INT((v# - r#) / 5)
LOOP
```

```
"C:\Users\Reactor1967\vcns\work\022003D.BAS"
REM ideal, keep v# - r# within a specific range.
REM the low end of that range(v2# - r2#) tells when to decrease
REM r# when decoding. The high end of the range(v2# - r2#) tells
REM when to increase r# when coding.
REM problem you can use (v1# - r1#) before coding to
REM code to (v2# - r2#) after coding. Now how to use
REM that to the advantage is the problem and might be
REM the solution.
v# = 1
r# = 1
spr# = 0
CLS
```

```
RANDOMIZE TIMER

DO

z = INT(RND * 2) + 0

dist# = (v# - r#)

v# = v# + (v# - r#) + z + 1

PRINT , z; v#; r#; v# - r#; spr#; flag; dist#

sv# = v#

sr# = r#

a$ = INKEY$

REM INPUT a$

IF a$ = "s" THEN STOP

IF v# - r# >= (spr# + 2) THEN flag = 1 ELSE flag = 0

IF flag = 1 THEN spr# = spr# + 2

IF flag = 1 THEN r# = r# + spr#

LOOP
```

```
"C:\Users\Reactor1967\vcns\work\022003C.BAS"

v# = 1

r# = 1

spr# = 0

RANDOMIZE TIMER

CLS
```

```
DO

z = INT(RND * 2) + 0

dist# = (v# - r#)

IF z = 0 THEN v# = v# + (v# - r# + 1)

IF z = 1 THEN v# = v# + (v# - r# + 2)

IF flag = 1 THEN PRINT , , v#; r#; v# - r#; spr#

IF flag = 0 THEN PRINT , , v#; r#; v# - r#

sv# = v#

sr# = r#

a$ = INKEY$

IF a$ = "s" THEN STOP

IF (spr# + 2) + r# < v# THEN flag = 1 ELSE flag = 0

IF flag = 1 THEN spr# = spr# + 2

IF flag = 1 THEN r# = r# + spr#

LOOP
```

```
"C:\Users\Reactor1967\vcns\work\022003B.BAS"

d# = 1

v# = 100

CLS

RANDOMIZE TIMER
```

```
DO
r# = (v# - d#)
IF (r# / 2) - INT(r# / 2) = 0 THEN d# = d# + 1
r# = (v# - d#)
z = INT(RND * 2) + 0
v# = v# + (v# - r#) + z + 1
PRINT , v#; r#; v# - r#; d#; v# - sv#; r# - sr#
sv# = v#
sr# = r#
a$ = INKEY$
REM input a$
IF a$ = "s" THEN STOP
d# = d# + 1
LOOP
```

```
"C:\Users\Reactor1967\vcns\work\022003A.BAS"
v1# = 100
v2# = 100
d# = 1
RANDOMIZE TIMER
CLS
```

```
DO
z1 = INT(RND * 2) + 0
r1# = v1# - d#
v1# = v1# + (v1# - r1#) + z1 + 1
z2 = INT(RND * 2) + 0
r2# = v2# - d#
v2# = v2# + (v2# - r2#) + z2 + 1
PRINT , , z1; v1#; r1#; d#; z2; v2#; r2#
a$ = INKEY$
IF a$ = "s" THEN STOP
d# = INT(RND * 100) + 1
LOOP
```

"C:\Users\Reactor1967\vcns\work\021903D.BAS"

```
d1# = 1: d2# = 2
v# = 100
CLS
DO
x2# = d2# - 2
x2# = x2# / 2
x1# = d1# - 1
x1# = x1# / 2
```

```
z = INT(RND * 2) + 0

r# = v# - x2#

v# = v# + (v# - r#) + z + 1

PRINT d1#; x1#; d2#; x2#, z; v#; r#; v# - r#

d1# = d1# + 2

d2# = d2# + 2

a$ = INKEY$

IF a$ = "s" THEN STOP

LOOP
```

"C:\Users\Reactor1967\vcns\work\021903C.BAS"

```
REM this program proves that you can code to what ever distance you
want

REM ending dist# = beginning (dist# * 2) + N

v# = 1273

r# = 574

CLS

RANDOMIZE TIMER

PRINT (v# - r#)

z = INT(RND * 2) + 0

v# = v# + (v# - r#) + z + 1
```

```
PRINT (v# - r#)

r# = v# - 5: REM 5 is beginning distance. Out come will be 11 or 12
always

z = INT(RND * 2) + 0

v# = v# + (v# - r#) + z + 1

PRINT v# - r#




"C:\Users\Reactor1967\vcns\work\021903B.BAS"

REM try to get some type of distance chart down then try to get some

REM type of pattern or equation for working that chart.

v# = 1

r# = 1

CLS

RANDOMIZE TIMER

DO

z = INT(RND * 2) + 0

bn# = (v# - r#)

v# = v# + (v# - r#) + z + 1

PRINT , , z; v#; r#; v# - r#; bn#; v# - sv#; r# - sr#

sv# = v#

sr# = r#

a$ = INKEY$
```

```
IF a$ = "s" THEN STOP
dist# = (v# - r#)
IF (dist# / 2) - INT(dist# / 2) = .5 THEN dist# = dist# - 1
dist# = dist# / 2
IF (dist# / 2) - INT(dist# / 2) = .5 THEN dist# = dist# - 1
r# = r# + dist#
LOOP
```

```
"C:\Users\Reactor1967\vcns\work\021903A.BAS"
v# = 100
r# = 100
d# = 0
CLS
RANDOMIZE TIMER
DO
z = INT(RND * 2) + 0
IF z = 0 THEN d2# = d2# + 1
IF z = 1 THEN d2# = d2# + 2
v# = v# + (v# - r#) + z + 1
PRINT z; v#; r#; v# - r#; v# - sv#; r# - sr#; d#; (r# - sr#) - d#
sv# = v#
sr# = r#
```

```
a$ = INKEY$
IF a$ = "s" THEN STOP
d# = d# + 1
r# = v# - d#
LOOP
```

```
"C:\Users\Reactor1967\vcns\work\021803A.BAS"
v# = 1000
dist# = 100
CLS
RANDOMIZE TIMER
DO
d# = dist#
IF (d# / 2) - INT(d# / 2) = .5 THEN d# = d# - 1
d# = d# / 2
d# = d# + 1
spr# = (dist# - d#)
PRINT z; dist#; d#; spr#; v#; (v# - dist#); (v# - dist#) - vf#; (v# - dist#) + spr#; ((v# - dist#) + spr#) - vf2#
vf2# = (v# - dist#) + spr#
vf# = (v# - dist#)
a$ = INKEY$
```

```
REM input a$

IF a$ = "s" THEN STOP

z = INT(RND * 2) + 0

v# = v# + d#

d# = (d# * 2)

IF z = 0 THEN d# = d# + 1

IF z = 1 THEN d# = d# + 2

IF z = 0 THEN v# = v# + 1

IF z = 1 THEN v# = v# + 2

dist# = d#


LOOP
```

```
"C:\Users\Reactor1967\vcns\work\021303C.BAS"

REM MY FIRST DIRECT MANUAL CONTROL OF THE SPEED# OF R#

REM try to establish a mathmatical relationshiop between

REM the speed# of v# and the speed# of r# so that when you

REM decode you can tell when and how much to subtract from

REM the speed of r#

REM ok we know this

REM 1. the distance v1# - r1# can be even or odd
```

```
REM 2. the distance v1# - r2# times 2 + N affects (v2# - r2#)

REM 3. so establish a known pattern for coding (r2# - r1#) + N up

REM 4. so establish a known pattern for coding (r2# - r1#) - N down

v# = 1

r# = 1

spr# = 0

RANDOMIZE TIMER

CLS

DO

z = INT(RND * 2) + 0

dist# = (v# - r#)

v# = v# + (v# - r#) + z + 1

PRINT z; v#; r#; v# - r#; dist#; v# - sv#; (v# - sv#) - x2#; spr#; spr# - x#

x# = spr#

x2# = (v# - sv#)

sv# = v#

a$ = INKEY$

REM INPUT a$

IF a$ = "s" THEN STOP

spr# = spr# + 10

gh1# = (v# - r#)

DO

gh2# = gh1# - spr#

REM IF gh2# = spr# THEN EXIT DO
```

```
IF (gh2# - 1) = spr# THEN EXIT DO

IF (gh2# - 2) = spr# THEN EXIT DO

spr# = spr# - 1

a$ = INKEY$

IF a$ = "s" THEN STOP

LOOP

r# = r# + spr#


LOOP
```

"C:\Users\Reactor1967\vcns\work\021303B.BAS"

```
REM MY FIRST DIRECT MANUAL CONTROL OF THE SPEED# OF R#

REM try to establish a mathmatical relationshiop between

REM the speed# of v# and the speed# of r# so that when you

REM decode you can tell when and how much to subtract from

REM the speed of r#

v# = 1

r# = 1

spr# = 0

RANDOMIZE TIMER

CLS
```

```
DO
z = INT(RND * 2) + 0
dist# = (v# - r#)
v# = v# + (v# - r#) + z + 1
PRINT z; v#; r#; v# - r#; dist#; v# - sv#; (v# - sv#) - x2#; spr#; spr# - x#
x# = spr#
x2# = (v# - sv#)
sv# = v#
a$ = INKEY$
REM INPUT a$
IF a$ = "s" THEN STOP
d1# = (v# - r#) - (spr# + 0)
d2# = (v# - r#) - (spr# + 2)
d3# = (v# - r#) - (spr# + 4)
flag = 0
DO
IF d3# >= (spr# + 4) THEN flag = 3
IF flag = 3 THEN EXIT DO
IF d2# >= (spr# + 2) THEN flag = 2
IF flag = 2 THEN EXIT DO
IF d1# >= (spr# + 0) THEN flag = 1
IF flag = 1 THEN EXIT DO
STOP
LOOP
```

```
IF flag = 3 THEN spr# = spr# + 4
IF flag = 2 THEN spr# = spr# + 2
IF flag = 1 THEN spr# = spr# + 0
r# = r# + spr#
LOOP
```

"C:\Users\Reactor1967\vcns\work\021303A.BAS"

```
REM In order to control this system you have to manually control the
REM speed of (v2# - r2#),v1# - r2#,v2# - v1#,r2# - r1#
REM to do so look at v2# - r2# and break that number into two things.
REM 0. (v1# - r2#) + (r2# - r1#) = (v2# - r2#)
REM 1. the next distance for v1# - r2#
REM 2. the next distance for r2# - r1#
REM v1# - r2# have to be in sprecific tolerences of the previous v1# - r2#
REM r2# - r1# have to be in specific tolerences of the previous r2# - r1#
REM the purpose of this program is to figure out those tolerences and
REM find a controlable system for doing the above mentioned things.
v# = 1
r# = 1
dist# = 0
CLS
RANDOMIZE TIMER
```

```
DO

test = (v# / 2) - INT(v# / 2)

IF test = .5 THEN z1 = 1 ELSE z1 = 0

z = INT(RND * 2) + 0

IF z1 = z THEN dist# = dist# + 2 ELSE dist# = dist# + 1

v# = v# + (v# - r#) + z + 1

PRINT v#; r#; v# - r#; (v# - r#) - x4#; dist#; dist# - x3#; v# - sv#; (v# -
sv#) - x2#; r# - sr#; (r# - sr#) - x1#

a$ = INKEY$

IF a$ = "s" THEN STOP

x1# = (r# - sr#)

x2# = (v# - sv#)

x3# = dist#

x4# = (v# - r#)

sv# = v#

sr# = r#

r# = v# - dist#

LOOP
```

```
"C:\Users\Reactor1967\vcns\work\021203A.BAS"

REM this program seems like it might work.

REM notice v# - (r# + increment) is positive when the r# changes
```

```
REM notice v# - (v# + increment) is negative when r# does not change.

REM for the above statments it seems true a good part of the time.

v# = 1

r# = 1

CLS

RANDOMIZE TIMER

DO

z = INT(RND * 2) + 0

v# = v# + (v# - r#) + z + 0

PRINT , , (v# - (r# + 50)); v#; r#; v# - r#; v# - sv#; r# - sr#

sv# = v#

sr# = r#

a$ = INKEY$

IF a$ = "s" THEN STOP

REM IF v# >= (r# + 50) AND (v# - (r# + 50) < 50) THEN r# = r# + 50

IF v# >= (r# + 50) THEN r# = r# + 50

LOOP
```

```
"C:\Users\Reactor1967\vcns\work\021103E.BAS"

REM try giving every vector a distance to multiply the base by

REM to get a value to add to r# for the next r#
```

```
REM So when you decode to the previous
REM vector you know the number to divide into the base by to get
REM a value to subtract from r# to get the previous r#
v# = 1
r# = 1
sv# = 1
sr# = 1
CLS
RANDOMIZE TIMER
DO
z = INT(RND * 2) + 0
test = (v# - r#) >= 100
r2# = r# + ((v# - r#) * 1)
dist# = (v# - r#)
v# = v# + ((v# - r#) * (2 - 1)) + z + 1
PRINT z; v#; r#; v# - r#; dist#; dist# - sd#; r# - sr#
sr# = r#
sd# = dist#
a$ = INKEY$
IF a$ = "s" THEN STOP
IF test = -1 THEN r# = r2#
IF (r# / 2) - INT(r# / 2) = 0 THEN r# = r# - 1
LOOP
```

```
"C:\Users\Reactor1967\vcns\work\021103D.BAS"

v# = 1

r# = 1

CLS

RANDOMIZE TIMER

DO

z = INT(RND * 10) + 0

test = (v# - r#) >= 100

r2# = r# + ((v# - r#) * 9)

IF z = 0 THEN v# = v# + ((v# - r#) * 9) + 1

IF z = 1 THEN v# = v# + ((v# - r#) * 9) + 2

IF z = 2 THEN v# = v# + ((v# - r#) * 9) + 3

IF z = 3 THEN v# = v# + ((v# - r#) * 9) + 4

IF z = 4 THEN v# = v# + ((v# - r#) * 9) + 5

IF z = 5 THEN v# = v# + ((v# - r#) * 9) + 6

IF z = 6 THEN v# = v# + ((v# - r#) * 9) + 7

IF z = 7 THEN v# = v# + ((v# - r#) * 9) + 8

IF z = 8 THEN v# = v# + ((v# - r#) * 9) + 9

IF z = 9 THEN v# = v# + ((v# - r#) * 9) + 10

PRINT z; v#; r#; v# - r#; v# - sv#; r# - sr#

sv# = v#

sr# = r#
```

```
a$ = INKEY$
IF a$ = "s" THEN STOP
IF test = -1 THEN r# = r2#
LOOP
```

"C:\Users\Reactor1967\vcns\work\021103C.BAS"

```
REM my first sucessful attempt at getting r# to increase in
REM other bases
v# = 1
r# = 1
CLS
RANDOMIZE TIMER
DO
z = INT(RND * 3) + 0
test = (v# - r#) >= 30
IF test = -1 THEN flag = 1 ELSE flag = 0
r2# = r# + ((v# - r#) * 2)
IF z = 0 THEN v# = v# + ((v# - r#) * 2) + 1
IF z = 1 THEN v# = v# + ((v# - r#) * 2) + 2
IF z = 2 THEN v# = v# + ((v# - r#) * 2) + 3
PRINT z; v#; r#; v# - r#; v# - sv#; r# - sr#
sv# = v#
```

```
sr# = r#
IF flag = 1 THEN r# = r2#
a$ = INKEY$
IF a$ = "s" THEN STOP
LOOP
```

"C:\Users\Reactor1967\vcns\work\021103B.BAS"

```
v# = 1
r# = 1
rm# = 0
CLS
RANDOMIZE TIMER
DO
z = INT(RND * 2) + 0
IF z = 0 THEN v# = v# + (v# - r# + 1)
IF z = 1 THEN v# = v# + (v# - r# + 2)
PRINT , , v#; r#; v# - r#; rm#
test = ((v# - r#) >= 50) AND ((v# + (v# - (r# + 50) + 1)) - (r# + 50) < 50)
IF test = -1 THEN flag = 1 ELSE flag = 0
IF flag = 1 THEN r# = r# + 50
IF flag = 1 THEN rm# = rm# + 50
sr# = r#
```

```
a$ = INKEY$
INPUT a$
IF a$ = "s" THEN STOP
LOOP
dist# = v# - r#
```

```
"C:\Users\Reactor1967\vcns\work\021103A.BAS"
v# = 25
r# = 1
CLS
RANDOMIZE TIMER
DO
z = INT(RND * 2) + 0
dist# = ABS(v# - r#)
IF z = 0 THEN v# = v# + (dist# + 1)
IF z = 1 THEN v# = v# + (dist# + 2)
PRINT , , v#; r#; v# - r#
IF v# - r# >= 50 THEN flag = 1 ELSE flag = 0
IF flag# = 1 THEN store1# = v#
IF flag# = 1 THEN store2# = r#
IF flag = 1 THEN v# = store2#
IF flag = 1 THEN r# = store1#
```

```
a$ = INKEY$

IF a$ = "s" THEN STOP

r# = r# + 2

LOOP
```

"C:\Users\Reactor1967\vcns\work\021003E.BAS"

```
v# = 100

r# = 1

CLS

RANDOMIZE TIMER

DO

d# = 0

z = INT(RND * 2) + 0

DO

d# = d# + 1

d1# = (d# * 2)

IF z = 0 THEN d1 = d1# + 1

IF z = 1 THEN d1# = d1# + 2: REM this is v2# - r2#

d2# = (v# - r#) - d#: REM this is the speed of r#

vt# = v# + d#

IF z = 0 THEN vt# = vt# + 1

IF z = 1 THEN vt# = v2# + 2: REM this is v2#
```

```
rt# = v2# - d1#: REM this is r2#

t1# = vt# - r#: REM this is v2# - r1#

test = d1# + (rt# - r#) = t1#

IF test = -1 THEN EXIT DO

a$ = INKEY$

IF a$ = "s" THEN STOP

LOOP UNTIL d# >= (v# - r#)

IF test = 0 THEN STOP

r# = v# - d#

IF z = 0 THEN v# = v# + (v# - r# + 1)

IF z = 1 THEN v# = v# + (v# - r# + 2)

PRINT , , v#; r#; v# - r#

a$ = INKEY$

IF a$ = "s" THEN STOP

LOOP
```

"C:\Users\Reactor1967\vcns\work\021003D.BAS"

```
v# = 2

r# = 2

dist# = 0

CLS
```

```
RANDOMIZE TIMER
DO
z = INT(RND * 2) + 0
IF z = 0 THEN v# = v# + (v# - r# + 1)
IF z = 1 THEN v# = v# + (v# - r# + 2)
PRINT z; v#; r#; v# - r#; dist#; v# - sr#
sv# = v#
sr# = r#
dist# = dist# + 2
r# = v# - dist#
a$ = INKEY$
IF a$ = "s" THEN STOP
LOOP
```

"C:\Users\Reactor1967\vcns\work\021003C.BAS"

```
v# = 50
dist# = 0
CLS
RANDOMIZE TIMER
DO
test = (v# / 2) - INT(v# / 2)
IF test = 0 THEN sz = 0 ELSE sz = 1
```

```
z = INT(RND * 2) + 0
IF z = sz THEN dist# = dist# + 2 ELSE dist# = dist# + 1
xr# = (v# - r#)
IF z = 0 THEN v# = v# + (v# - r# + 1)
IF z = 1 THEN v# = v# + (v# - r# + 2)
REM PRINT z; v#; r#; v# - r#; dist#; v# - sr#
PRINT z; v#; r#; v# - r#; xr#; dist#; v# - sr#; v# - sv#; r# - sr#
REM PRINT z; v#; r#
sv# = v#
sr# = r#
r# = v# - dist#
a$ = INKEY$
IF a$ = "s" THEN STOP
LOOP
```

```
"C:\Users\Reactor1967\vcns\work\021003B.BAS"
v# = 1
r# = 1
z = 0
CLS
DO
test = (v# / 2) - INT(v# / 2)
```

```
IF test = 0 THEN sz = 0

IF test = .5 THEN sz = 1

z = INT(RND * 2) + 0

IF sz = z THEN dist# = dist# + 2 ELSE dist# = dist# + 1

xr# = (v# - r#)

IF z = 0 THEN v# = v# + (v# - r# + 1)

IF z = 1 THEN v# = v# + (v# - r# + 2)

PRINT z; v#; r#; v# - r#; dist#; xr#; v# - sv#; r# - sr#

a$ = INKEY$

REM INPUT a$

IF a$ = "s" THEN STOP

sv# = v#: sr# = r#

r# = v# - dist#

REM r# = r# + (xr# - 1)

LOOP
```

```
"C:\Users\Reactor1967\vcns\work\021003A.BAS"

v# = 1

r# = 1

z = 1

dist# = 0

spr# = 0
```

```
CLS

DO

sz# = z

z = INT(RND * 2) + 0

IF sz = z THEN dist# = dist# + 2 ELSE dist# = dist# + 1

IF z = 0 THEN v# = v# + (v# - r# + 1)

IF z = 1 THEN v# = v# + (v# - r# + 2)

PRINT , , v#; r#; v# - r#; dist#; spr#

a$ = INKEY$

INPUT a$

IF a$ = "s" THEN STOP

test2 = (dist# - spr# = 1)

test3 = (dist# - spr# = 2)

test4 = (dist# - spr# = 3)

test5 = (dist# - spr# = 4)

test6 = (dist# - spr# >= 5)

IF test2 = -1 THEN spr# = spr# + 0

IF test3 = -1 THEN spr# = spr# + 2

IF test4 = -1 THEN spr# = spr# + 7

IF test5 = -1 THEN spr# = spr# + 8

r# = r# + spr#


LOOP
```

```
"C:\Users\Reactor1967\vcns\work\020903E.BAS"
REM speed of R# is propertional to the dist# used to code v#
REM when the dist# you use to code v# increases or decrease so does
REM your speed of r#. So use speed of r# in relation to dist# and
REM see if it works.
REM so if dist# - speed of r# => z then speed of sr# = sr# + y
CLS
RANDOMIZE TIMER
v# = 1
r# = 1
DO
z = INT(RND * 2) + 0
test = (v# / 2) - INT(v# / 2)
test1 = (z = 0) AND (test = 0)
test2 = (z = 0) AND (test = .5)
test3 = (z = 1) AND (test = 0)
test4 = (z = 1) AND (test = .5)
IF test1 = -1 THEN dist# = dist# + 2
IF test2 = -1 THEN dist# = dist# + 1
IF test3 = -1 THEN dist# = dist# + 1
IF test4 = -1 THEN dist# = dist# + 2
xr# = (v# - r#)
```

```
IF z = 0 THEN v# = v# + (v# - r# + 1)

IF z = 1 THEN v# = v# + (v# - r# + 2)

PRINT z; v#; r#; v# - r#; dist#; xr#; v# - sv#; r# - sr#

a$ = INKEY$

REM INPUT a$

IF a$ = "s" THEN STOP

sv# = v#

sr# = r#

r# = v# - dist#

LOOP
```

```
"C:\Users\Reactor1967\vcns\work\020903D.BAS"

v# = 1

r# = 1

CLS

RANDOMIZE TIMER

PRINT , , v#; r#; v# - r#

dist# = 0

DO

count = count + 1

z = INT(RND * 2) + 0

IF z = 0 THEN v# = v# + (v# - r# + 1)
```

```
IF z = 1 THEN v# = v# + (v# - r# + 2)

IF z = 0 THEN dist# = dist# + 1

IF z = 1 THEN dist# = dist# + 2

x# = dist#

IF z = 0 THEN x# = x# - 1

IF z = 1 THEN x# = x# - 2

d# = (v# - r#) - x#

r# = v# - d#

REM r# = r# + x#

PRINT z; v#; r#; v# - r#; v# - sv#; r# - sr#; dist#; count

a$ = INKEY$

REM INPUT a$

IF a$ = "s" THEN STOP

IF a$ = "d" THEN EXIT DO

sv# = v#

sr# = r#

LOOP

PRINT "--------------------------------------------"
```

"C:\Users\Reactor1967\vcns\work\020903C.BAS"

```
v# = 1

r# = 1
```

```
dist# = 0
CLS
DO
z = INT(RND * 2) + 0
IF z = 0 THEN v# = v# + (v# - r# + 1)
IF z = 1 THEN v# = v# + (v# - r# + 2)
PRINT z; v#; r#; v# - r#; d#; v# - sv#; r# - sr#; dist#
a$ = INKEY$
REM INPUT a$
IF a$ = "s" THEN STOP
sr# = r#
sv# = v#
IF z = 0 THEN dist# = dist# + 1
IF z = 1 THEN dist# = dist# + 2
x# = dist#
IF z = 0 THEN x# = x# - 1
IF z = 1 THEN x# = x# - 2
d# = (v# - r#) - x#
r# = v# - d#
LOOP
```

"C:\Users\Reactor1967\vcns\work\020903B.BAS"

```
v# = 1
r# = 1
CLS
RANDOMIZE TIMER
PRINT , , v#; r#; v# - r#
dist# = 0
DO
count = count + 1
z = INT(RND * 2) + 0
IF z = 0 THEN v# = v# + (v# - r# + 1)
IF z = 1 THEN v# = v# + (v# - r# + 2)
IF z = 0 THEN dist# = dist# + 1
IF z = 1 THEN dist# = dist# + 2
x# = dist#
IF z = 0 THEN x# = x# - 1
IF z = 1 THEN x# = x# - 2
r# = r# + x#
PRINT z; v#; r#; v# - r#; v# - sv#; r# - sr#; dist#; count
a$ = INKEY$
REM INPUT a$
IF a$ = "s" THEN STOP
IF a$ = "d" THEN EXIT DO
sv# = v#
sr# = r#
```

```
LOOP

PRINT "------------------------------------------"




"C:\Users\Reactor1967\vcns\work\020903A.BAS"

v# = 1

r# = 1

CLS

RANDOMIZE TIMER

count = 0

PRINT , , v#; r#; v# - r#

dist# = 0

DO

count = count + 1

z = INT(RND * 2) + 0

IF z = 0 THEN dist# = dist# + 1

IF z = 1 THEN dist# = dist# + 2

rx# = (v# - r#)

IF z = 0 THEN v# = v# + (v# - r# + 2)

IF z = 1 THEN v# = v# + (v# - r# + 4)

PRINT z; v#; r#; v# - r#; v# - sv#; r# - sr#; rx#; dist#

a$ = INKEY$

REM INPUT a$
```

```
IF a$ = "s" THEN STOP

IF a$ = "d" THEN EXIT DO

sv# = v#

sr# = r#

r# = r# + dist#

LOOP

PRINT "--------------------------------------------"

x# = v#

test1 = ((r# + 2) / 4) - INT((r# + 2) / 4)

test2 = ((r# + 4) / 4) - INT((r# + 4) / 4)

test3 = (v# / 4) - INT(v# / 4)

test4 = (test1 = test3)

test5 = (test2 = test3)

IF test4 = -1 THEN z = 0

IF test5 = -1 THEN z = 1

PRINT z; v#; r#; v# - r#; dist#

a$ = INKEY$

REM INPUT a$

IF a$ = "s" THEN STOP
```

```
"C:\Users\Reactor1967\vcns\work\020803C.BAS"

v# = 1
```

```
r# = 1

d1# = 1

RANDOMIZE TIMER

CLS

z = 0

count = 0

DO

count = count + 1

REM z = INT(RND * 2) + 0

IF z = 0 THEN v# = v# + (v# - r# + 1)

IF z = 1 THEN v# = v# + (v# - r# + 2)

REM PRINT , , v#; r#; v# - r#; v# - sv#; r# - sr#

REM PRINT , , v#; r#; v# - r#; d1#; count

PRINT , , z; v#; r#; count; v# - sv#; v# - r#

sr# = r#

sv# = v#

a$ = INKEY$

REM INPUT a$

IF a$ = "s" THEN STOP

IF a$ = "d" THEN EXIT DO

REM IF d1# + 100 >= 99999 THEN EXIT DO

test = v# >= d1# + 99

IF test = -1 THEN r# = r# + 100

IF test = -1 THEN d1# = d1# + 100
```

```
IF test = -1 THEN z = 1 ELSE z = 0
LOOP
INPUT a$
IF a$ = "s" THEN STOP
DO
test = (v# / 2) - INT(v# / 2) = .5
REM PRINT , , v#; r#; v# - r#; d1#; count
PRINT , , v#; r#; count
a$ = INKEY$
REM INPUT a$
IF a$ = "s" THEN STOP
IF v# <= 1 THEN EXIT DO
dist# = v# - r#
IF (dist# / 2) - INT(dist# / 2) = .5 THEN dist# = dist# - 1
dist# = dist# / 2
dist# = dist# + 1
v# = v# - dist#
IF test = -1 THEN r# = r# - 100
IF test = -1 THEN d1# = d1# - 100
count = count - 1
LOOP
```

"C:\Users\Reactor1967\vcns\work\020803B.BAS"

```
CLS

RANDOMIZE TIMER

DO

d# = INT(RND * 256) + 100

x# = d#

IF (x# / 2) - INT(x# / 2) = .5 THEN x# = x# - 1

x# = x# / 2

x# = x# + 1

spr# = d# - x#

PRINT , , d#; x#; spr#; x# - spr#

PRINT " "

a$ = INKEY$

IF a$ = "s" THEN STOP

LOOP
```

"C:\Users\Reactor1967\vcns\work\020803A.BAS"

```
REM process for finding x# which is the distance you code when. you

REM can divide x# by 4 to get a fractional value(remainder) which tells
```

REM you how much to subtract from x# to get your speed of r#. Take your

REM speed and subtract it from your current r# to get your previous r#

d# = 247

CLS

DO

x# = d#

x# = x# - 1

x# = x# / 2

x# = x# - 1: REM -1 can be anything. this determines the distance between d# & x#

spr# = d# - x#

PRINT d#; x#; spr#; x# - spr#; x# / 4

INPUT a$

IF a$ = "s" THEN STOP

d# = d# + 2

LOOP

"C:\Users\Reactor1967\vcns\work\020703B.BAS"

v# = 10

r# = 4

CLS

RANDOMIZE TIMER

```
DO
z = INT(RND * 2) + 0
dist# = (v# - r#)
test3# = (dist# * 2)
IF z = 0 THEN test3# = test3# + 1
IF z = 1 THEN test3# = test3# + 2
d# = 0
DO
d# = d# + 1
x# = (dist# - d#)
REM if (dist# = even) (speed of r = even)
REM if (dist# = odd) (speed of r = odd)
REM know if dist# = speed of r or if speed of r = dist# - 1
PRINT dist#; d#; x#; test3#
a$ = INKEY$
INPUT a$
IF a$ = "s" THEN STOP
LOOP
REM r# = r# + x#
r# = v# - d#
dist# = v# - r#
IF z = 0 THEN v# = v# + (v# - r# + 1)
IF z = 1 THEN v# = v# + (v# - r# + 2)
PRINT , , z; v#; r#; v# - r#; dist#; v# - sv#; r# - sr#
```

```
sv# = v#

sr# = r#

a$ = INKEY$

IF a$ = "s" THEN STOP

LOOP
```

"C:\Users\Reactor1967\vcns\work\020703A.BAS"

```
v# = 10

r# = 4

CLS

RANDOMIZE TIMER

DO

z = INT(RND * 2) + 0

dist# = (v# - r#)

d# = 0

DO

d# = d# + 1

x# = (dist# - d#)

REM test3 = (x# / 2) - INT(x# / 2) = 0

REM test1 = (x# = d#) AND (test3 = -1)

test1 = (x# = d#)

REM test2 = (x# + 1 = d#) AND (test3 = -1)
```

```
test2 = (x# + 1 = d#)

IF test1 = -1 THEN EXIT DO

IF test2 = -1 THEN EXIT DO

a$ = INKEY$

IF a$ = "s" THEN STOP

LOOP

REM r# = r# + x#

r# = v# - d#

dist# = v# - r#

IF z = 0 THEN v# = v# + (v# - r# + 1)

IF z = 1 THEN v# = v# + (v# - r# + 2)

PRINT , , z; v#; r#; v# - r#; dist#; v# - sv#; r# - sr#

sv# = v#

sr# = r#

a$ = INKEY$

IF a$ = "s" THEN STOP

LOOP




"C:\Users\Reactor1967\vcns\work\020603E.BAS"

v# = 10

r# = 4

CLS
```

```
RANDOMIZE TIMER
DO
z = INT(RND * 2) + 0
dist# = (v# - r#)
d# = 0
DO
d# = d# + 1
x# = (dist# - d#)
test1 = (x# = d#)
test2 = (x# - 1 = d#)
IF test1 = -1 THEN EXIT DO
IF test2 = -1 THEN EXIT DO
a$ = INKEY$
IF a$ = "s" THEN STOP
LOOP
r# = r# + x#
IF z = 0 THEN v# = v# + (v# - r# + 1)
IF z = 1 THEN v# = v# + (v# - r# + 2)
PRINT , , v#; r#; v# - r#; v# - sv#; r# - sr#
a$ = INKEY$
IF a$ = "s" THEN STOP
LOOP
```

```
"C:\Users\Reactor1967\vcns\work\020603D.BAS"

v# = 1

r# = 1

CLS

RANDOMIZE TIMER

DO

z = INT(RND * 2) + 0

dist# = v# - r#

IF z = 0 THEN v# = v# + (v# - r# + 1)

IF z = 1 THEN v# = v# + (v# - r# + 2)

PRINT z; v#; r#; dist#; dist# - sd#; v# - r#; (v# - r#) - svr#; v# - sv#; (v# -
sv#) - x1#; r# - sr#; (r# - sr#) - x2#

x1# = (v# - sv#)

x2# = (r# - sr#)

x3# = (dist# - sd#)

sd# = dist#

sv# = v#

sr# = r#

svr# = (v# - r#)

REM dist# = (r# - svr#)

REM IF (dist# / 2) - INT(dist# / 2) = .5 THEN dist# = dist# - 1

REM r# = v# - dist#

r# = r# + dist#
```

```
a$ = INKEY$
REM INPUT a$
IF a$ = "s" THEN STOP
LOOP
```

"C:\Users\Reactor1967\vcns\work\020603C.BAS"

```
v# = 1
r# = 1
RANDOMIZE TIMER
count = 0
DO
count = count + 1
z = INT(RND * 2) + 0
dist# = v# - r#
IF z = 0 THEN v# = v# + (v# - r# + 1)
IF z = 1 THEN v# = v# + (v# - r# + 2)
PRINT , , z; v#; r#; v# - r#; count; r# - sr#
sr# = r#
a$ = INKEY$
IF a$ = "s" THEN STOP
x# = (count / 2) - INT(count / 2)
IF x# = .5 THEN r# = r# + (dist# * 2)
```

```
LOOP

"C:\Users\Reactor1967\vcns\work\020603B.BAS"
REM Next time around do
REM do if (v# - r#) => 499 then r# = r# + 500 coding up
REM decoding put the r2# = ( v# + (v# - r#)) above v# and
REM use r2# to decode with resprect to the true r#
v# = 1
r# = 1
CLS
RANDOMIZE TIMER
count = 0
DO
count = count + 1
z = INT(RND * 2) + 0
IF (count / 2) - INT(count / 2) = .5 THEN z = 1
IF z = 0 THEN v# = v# + (v# - r# + 1)
IF z = 1 THEN v# = v# + (v# - r# + 2)
REM IF v# >= r# + 499 THEN flag = 1 ELSE flag = 0
REM PRINT , , flag; v#; r#; v# - r#
PRINT z; v#; r#; v# - r#; count
```

```
a$ = INKEY$
REM INPUT a$
IF a$ = "d" THEN EXIT DO
IF a$ = "s" THEN STOP
REM IF flag = 1 THEN r# = r# + 500
dist# = v# - r#
test = ((dist# / 2) - INT(dist# / 2) = 0)
IF test = -1 THEN r# = r# + ((dist# / 2) * 1)
REM IF test = -1 THEN r# = r# + ((dist# / 2) * 2)
LOOP UNTIL v# + (v# - r# + 1) >= 999999999999999#
```

```
"C:\Users\Reactor1967\vcns\work\020603A.BAS"
REM Next time around do
REM do if (v# - r#) => 499 then r# = r# + 500 coding up
REM decoding put the r2# = ( v# + (v# - r#)) above v# and
REM use r2# to decode with resprect to the true r#
v# = 1
r# = 1
rx# = 500
CLS
RANDOMIZE TIMER
DO
```

```
z = INT(RND * 2) + 0

dist# = v# - r#

REM IF (v# + (v# - r#)) >= rx# THEN z = 1

IF z = 0 THEN v# = v# + (v# - r# + 1)

IF z = 1 THEN v# = v# + (v# - r# + 2)

PRINT , , rx#; v#; rx# - 500; r#

a$ = INKEY$

REM INPUT a$

IF a$ = "s" THEN STOP

IF a$ = "d" THEN EXIT DO

sv# = v#

sr# = r#

REM IF (v# - r#) >= 499 THEN r# = r# + 500

IF v# >= rx# THEN r# = r# + 500

IF v# >= rx# THEN rx# = rx# + 500

LOOP
```

```
"C:\Users\Reactor1967\vcns\work\020503J.BAS"

v# = 50

r# = 49

CLS

RANDOMIZE TIMER
```

```
DO

z = INT(RND * 2) + 0

dist# = (v# - r#)

d# = 1

DO

x# = d# * 2

IF z = 0 THEN x# = x# + 1

IF z = 1 THEN x# = x# + 2

test = (x# - dist# >= 4) AND (((v# - d#) / 2) - INT((v# - d#) / 2) = .5)

IF test = -1 THEN EXIT DO

d# = d# + 1

a$ = INKEY$

IF a$ = "s" THEN STOP

LOOP

r# = v# - d#

IF z = 0 THEN v# = v# + (v# - r# + 1)

IF z = 1 THEN v# = v# + (v# - r# + 2)

PRINT z; v#; r#; v# - r#; (v# - r#) - X1#; dist#; dist# - x5#; v# - sv#; (v# -
sv#) - x2#; r# - sr#; (r# - sr#) - x3#

X1# = (v# - r#)

x2# = (v# - sv#)

x3# = (r# - sr#)

x5# = dist#

sv# = v#
```

```
sr# = r#

a$ = INKEY$

IF a$ = "s" THEN STOP

LOOP
```

"C:\Users\Reactor1967\vcns\work\020503I.BAS"

```
v# = 1

r# = 1

dist# = 0

CLS

RANDOMIZE TIMER

DO

z = INT(RND * 2) + 0

test = (v# / 2) - INT(v# / 2)

IF test = .5 THEN test = 1

IF z = test THEN dist# = dist# + 2 ELSE dist# = dist# + 1

xt# = (v# - r#)

IF z = 0 THEN v# = v# + (v# - r# + 1)

IF z = 1 THEN v# = v# + (v# - r# + 2)

PRINT z; v#; r#; v# - r#; (v# - r#) - X1#; xt#; (xt# - x4#); dist#; dist# -
x5#; v# - sv#; (v# - sv#) - x2#; r# - sr#; (r# - sr#) - x3#
```

```
X1# = (v# - r#)

x2# = (v# - sv#)

x3# = (r# - sr#)

x4# = xt#

x5# = dist#

sv# = v#

sr# = r#

r# = v# - dist#

a$ = INKEY$

IF a$ = "s" THEN STOP

LOOP
```

"C:\Users\Reactor1967\vcns\work\020503H.BAS"

```
v# = 238

r# = 117

xo = 1

RANDOMIZE TIMER

CLS

DO

z = INT(RND * 2) + 0

dist# = v# - r#
```

```
IF z = xo THEN dist# = dist# + 12

IF z <> xo THEN dist# = dist# + 13

IF (dist# / 2) - INT(dist# / 2) = 0 THEN sb# = 2

IF (dist# / 2) - INT(dist# / 2) = .5 THEN sb# = 1

dist# = dist# - sb#

dist# = dist# / 2

r# = v# - dist#

IF z = 0 THEN v# = v# + (v# - r# + 1)

IF z = 1 THEN v# = v# + (v# - r# + 2)

PRINT , z; v#; r#; v# - r#; dist#; v# - sv#; r# - sr#

xo = z

sv# = v#

sr# = r#

a$ = INKEY$

IF a$ = "s" THEN STOP

LOOP
```

```
"C:\Users\Reactor1967\vcns\work\020503G.BAS"

v# = 238

r# = 117

RANDOMIZE TIMER
```

```
CLS

DO

z = INT(RND * 2) + 0

dist# = v# - r#

IF z = 0 THEN v# = v# + (v# - r# + 1)

IF z = 1 THEN v# = v# + (v# - r# + 2)

PRINT , z; v#; r#; v# - r#; dist#; v# - sv#; r# - sr#

sv# = v#

sr# = r#

REM IF z = 0 THEN dist# = dist# - 1

REM IF z = 1 THEN dist# = dist# - 2

IF (dist# / 2) - INT(dist# / 2) = .5 THEN dist# = dist# - 1

r# = r# + dist#

a$ = INKEY$

INPUT a$

IF a$ = "s" THEN STOP

LOOP
```

"C:\Users\Reactor1967\vcns\work\020503F.BAS"

```
v# = 50

r# = 45

sr# = 40
```

```
RANDOMIZE TIMER

CLS

DO

z = INT(RND * 2) + 0

IF z = 0 THEN v# = v# + (v# - r# + 1)

IF z = 1 THEN v# = v# + (v# - r# + 2)

PRINT , z; v#; r#; v# - r#; dist#; v# - sv#; r# - sr#

dist# = (v# - sv#)

sv# = v#

sr# = r#

r# = v# - dist#

REM IF (r# / 2) - INT(r# / 2) = 0 THEN r# = r# - 1

a$ = INKEY$

INPUT a$

IF a$ = "s" THEN STOP

LOOP
```

```
"C:\Users\Reactor1967\vcns\work\020503E.BAS"

v# = 50

r# = 45

dist# = 5

RANDOMIZE TIMER
```

```
CLS

DO

z = INT(RND * 2) + 0

IF z = 0 THEN v# = v# + (v# - r# + 1)

IF z = 1 THEN v# = v# + (v# - r# + 2)

PRINT , z; v#; r#; v# - r#; dist#; v# - sv#; r# - sr#

sv# = v#

sr# = r#

r# = r# + dist#

IF (r# / 2) - INT(r# / 2) = 0 THEN r# = r# - 1

dist# = (v# - r#)

a$ = INKEY$

REM INPUT a$

IF a$ = "s" THEN STOP

LOOP
```

```
"C:\Users\Reactor1967\vcns\work\020503D.BAS"

DECLARE SUB cal (stat!, d2#, x1#, s1#, s2#)

REM successful program to demonstrat climming

REM the ladder.bas program earlier written.

v# = 50

r# = 47
```

```
CLS
RANDOMIZE TIMER
DO
z = INT(RND * 2) + 0
x1# = (v# - r#)
test = (v# / 2) - INT(v# / 2)
test1 = (test = 0 AND z = 0)
test2 = (test = 0 AND z = 1)
test3 = (test = .5 AND z = 0)
test4 = (test = .5 AND z = 1)
IF test1 = -1 THEN stat = 1
IF test2 = -1 THEN stat = 2
IF test3 = -1 THEN stat = 3
IF test4 = -1 THEN stat = 4
d2# = 0
CALL cal(stat, d2#, x1#, s1#, s2#)
s1# = d2#: s2# = stat
test = (d2# / 2) - INT(d2# / 2)
IF test = 0 THEN d2# = d2# - 2
IF test = .5 THEN d2# = d2# - 1
d2# = d2# / 2
r# = v# - d2#
IF z = 0 THEN v# = v# + (v# - r# + 1)
IF z = 1 THEN v# = v# + (v# - r# + 2)
```

```
PRINT z; v#; r#; v# - r#; (v# - r#) - drv#; d2#; v# - sv#; r# - sr#; stat
drv# = (v# - r#)
sv# = v#
sr# = r#
a$ = INKEY$
REM INPUT a$
IF a$ = "s" THEN STOP
LOOP

SUB cal (stat, d2#, x1#, s1#, s2#)
d# = 2
status = 1
REM IF x1# > 3 THEN d# = s1#
REM IF x1# > 3 THEN status = s2#
DO
d# = d# + 1
REM --------------
REM do stuff here
test = (stat = status) AND (d# > x1#)
IF test = -1 THEN EXIT DO
REM --------------
IF status = 4 THEN status = 0
status = status + 1
a$ = INKEY$
```

```
IF a$ = "s" THEN STOP
LOOP
d2# = d#
END SUB
```

"C:\Users\Reactor1967\vcns\work\020503C.BAS"

```
REM this program is a failed test. It did not work as intended
REM  lesson learned. Devise a test so that distance can
REM be coded from 0 to 0, 0 to 1, 1 to 0, and 1 to 1
REM be able to test the distance and work it backups for decode.
v# = 50
r# = 47
CLS
RANDOMIZE TIMER
DO
z = INT(RND * 2) + 0
dist# = v# - r#
test1:
dist# = dist# + 2
x# = dist#
IF (x# / 2) - INT(x# / 2) = 0 THEN x# = x# - 2
IF (x# / 2) - INT(x# / 2) = .5 THEN x# = x# - 3
```

```
a$ = INKEY$: IF a$ = "s" THEN STOP

IF (x# / 4) - INT(x# / 4) <> 0 THEN GOTO test1:

x# = dist#

IF (x# / 2) - INT(x# / 2) = 0 THEN x# = x# - 2

IF (x# / 2) - INT(x# / 2) = .5 THEN x# = x# - 1

x# = x# / 2

x# = (x# * 2)

IF z = 0 THEN x# = x# + 1

IF z = 1 THEN x# = x# + 2

IF (x# / 2) - INT(x# / 2) = 0 THEN x# = x# - 2

IF (x# / 2) - INT(x# / 2) = .5 THEN x# = x# - 3

a$ = INKEY$: IF a$ = "s" THEN STOP

IF (x# / 4) - INT(x# / 4) <> 0 THEN GOTO test1:

r# = v# - dist#

IF z = 0 THEN v# = v# + (v# - r# + 1)

IF z = 1 THEN v# = v# + (v# - r# + 2)

PRINT , , v#; r#; v# - r#; v# - sv#; r# - sr#

a$ = INKEY$

IF a$ = "s" THEN STOP

LOOP
```

```
"C:\Users\Reactor1967\vcns\work\020503B.BAS"
REM BUILDING A LADDER. Get a process that climbs up or down
REM and you can use it go code with and decode with.
REM so looking at v - r if it is odd subtracit it from 3
REM if v - r is even subtract it from 2
REM then divide the product by 4. the remainder must be 0
REM use this to test coding and coding down.
v# = 50
r# = 43
CLS
RANDOMIZE TIMER
DO
z = INT(RND * 2) + 0
dist# = (v# - r#)
dist# = dist# + 2
x# = dist#
test = (dist# / 2) - INT(dist# / 2)
IF test = 0 THEN x# = x# - 2
IF test = .5 THEN x# = x# - 3
IF (x# / 4) - INT(x# / 4) <> 0 THEN dist# = dist# + 2
IF test = 0 THEN dist# = dist# - 2
IF test = .5 THEN dist# = dist# - 1
dist# = dist# / 2
r# = v# - dist#
```

```
IF z = 0 THEN v# = v# + (v# - r# + 1)

IF z = 1 THEN v# = v# + (v# - r# + 2)

PRINT v#; r#; v# - r#; (v# - r#) - x#; dist#; v# - sv#; r# - sr#

x# = v# - r#

sv# = v#

sr# = r#

a$ = INKEY$

IF a$ = "s" THEN STOP

LOOP
```

```
"C:\Users\Reactor1967\vcns\work\020503A.BAS"

REM this program is used for attempting to calculate rate of change

REM for v# - r#. The goal here is to use this for

REM binary 0 to a binary 0 rate of change = x1

REM binary 0 to a binary 1 rate of change = x2

REM binary 1 to a binary 0 rate of change = x3

REM binary 1 to a binary 1 rate of change = x4

REM If I can get a specific rate of change for

REM v# - r# for (x1 to x4) then it might be possible

REM to get a decode

redo:

CLS
```

```
INPUT "Enter starting (v# - r#)"; lb1
INPUT "Enter desired rate of change"; rate
d# = 2
d2# = 3
count = 0
DO
PRINT , , d#; count; x#; d2#; count2; x2#
REM INPUT a$
REM IF a$ = "s" THEN STOP
test1 = ((d# - lb1) >= rate) OR ((d2# - lb1) >= rate)
IF test1 = -1 THEN EXIT DO
count = 0
count2 = 0
DO
count = count + 1
d# = d# + 2
x# = d# - 2
x# = x# / 2
IF (x# / 2) - INT(x# / 2) = .5 THEN EXIT DO
a$ = INKEY$
IF a$ = "s" THEN STOP
LOOP
a$ = INKEY$
IF a$ = "s" THEN STOP
```

```
DO
count2 = count2 + 1
d2# = d2# + 2
x2# = d2# - 1
x2# = x2# / 2
IF (x2# / 2) - INT(x2# / 2) = 0 THEN EXIT DO
a$ = INKEY$
IF a$ = "s" THEN STOP
LOOP
a$ = INKEY$
IF a$ = "s" THEN STOP
LOOP
PRINT "----------------------------------------------------"
PRINT lb1; rate
PRINT d# - lb1; d2# - lb1
PRINT , , d#; count; x#; d2#; count2; x2#
INPUT a$
IF a$ = "s" THEN STOP
GOTO redo:



"C:\Users\Reactor1967\vcns\work\020503.BAS"
v# = 238
```

```
r# = 117

RANDOMIZE TIMER

CLS

DO

z = INT(RND * 2) + 0

dist# = v# - r#

IF z = 0 THEN v# = v# + (v# - r# + 1)

IF z = 1 THEN v# = v# + (v# - r# + 2)

PRINT , z; v#; r#; v# - r#; dist#; v# - sv#; r# - sr#

sv# = v#

sr# = r#

REM IF z = 0 THEN dist# = dist# - 1

REM IF z = 1 THEN dist# = dist# - 2

IF (dist# / 2) - INT(dist# / 2) = .5 THEN dist# = dist# - 1

r# = r# + dist#

a$ = INKEY$

INPUT a$

IF a$ = "s" THEN STOP

LOOP
```

"C:\Users\Reactor1967\vcns\work\020403B.BAS"

```
v# = 50
```

```
sz = 0

CLS

RANDOMIZE TIMER

DO

z = INT(RND * 2) + 0

IF sz = 0 THEN d# = 4

IF sz = 1 THEN d# = 5

sz = z

r# = v# - d#

IF z = 0 THEN v# = v# + (v# - r# + 1)

IF z = 1 THEN v# = v# + (v# - r# + 2)

PRINT , , z; v#; r#; v# - r#; (v# - r#) - x#; d#; v# - sv#; r# - sr#

a$ = INKEY$

IF a$ = "s" THEN STOP

sv# = v#

sr# = r#

x# = (v# - r#)

LOOP
```

```
"C:\Users\Reactor1967\vcns\work\020403A.BAS"

sv# = 50

d# = 0
```

```
CLS

DO

z = INT(RND * 2) + 0

IF z = 0 THEN d# = d# + 1

IF z = 1 THEN d# = d# + 2

r# = v# - d#

REM IF (r# / 2) - INT(r# / 2) = 0 THEN r# = r# - 1

IF z = 0 THEN v# = v# + (v# - r# + 1)

IF z = 1 THEN v# = v# + (v# - r# + 2)

PRINT z; v#; r#; v# - r#; (v# - r#) - x2#; d#; v# - sv#; r# - sr#; (r# - sr#) - x#

x# = r# - sr#

x2# = (v# - r#)

sv# = v#

sr# = r#

a$ = INKEY$

REM INPUT a$

IF a$ = "s" THEN STOP

LOOP
```

"C:\Users\Reactor1967\vcns\work\020309A.BAS"

```
v# = 1
```

```
r# = 1

CLS

RANDOMIZE TIMER

count = 0

PRINT , , v#; r#; v# - r#

dist# = 0

DO

count = count + 1

z = INT(RND * 2) + 0

IF z = 0 THEN dist# = dist# + 1

IF z = 1 THEN dist# = dist# + 2

IF z = 0 THEN v# = v# + (v# - r# + 2)

IF z = 1 THEN v# = v# + (v# - r# + 4)

PRINT z; v#; r#; v# - r#; v# - sv#; r# - sr#; dist#

decode# = (r# - sr#)

sv# = v#

sr# = r#

a$ = INKEY$

REM INPUT a$

IF a$ = "s" THEN STOP

r# = r# + dist#

LOOP
```

```
"C:\Users\Reactor1967\vcns\work\020308A.BAS"

v# = 1

r# = 1

d1# = 1

d2# = 99999

RANDOMIZE TIMER

CLS

DO

z = INT(RND * 2) + 0

IF z = 0 THEN v# = v# + (v# - r# + 1)

IF z = 1 THEN v# = v# + (v# - r# + 2)

PRINT , , v#; r#; v# - r#; v# - sv#; r# - sr#

REM PRINT , , v#; r#; d1#; d2#

sv# = v#

sr# = r#

a$ = INKEY$

REM INPUT a$

IF a$ = "s" THEN STOP

IF a$ = "d" THEN EXIT DO

test = (v# >= d1# + 99) AND (99999 - v# <= d2# - 99)

IF test = -1 THEN r# = r# + 100

IF test = -1 THEN d1# = d1# + 100

IF test = -1 THEN d2# = d2# - 100
```

```
LOOP
```

"C:\Users\Reactor1967\vcns\work\020307A.BAS"

```
v# = 1
r# = 1
CLS
RANDOMIZE TIMER
DO
z = INT(RND * 2) + 0
dist# = (v# - r#)
IF z = 0 THEN v# = v# + (v# - r# + 1)
IF z = 1 THEN v# = v# + (v# - r# + 2)
PRINT , , z; v#; r#; v# - r#; dist#; (v# - r#) - x1#
x1# = (v# - r#)
a$ = INKEY$
IF a$ = "s" THEN STOP
r# = r# + dist#
LOOP
```

"C:\Users\Reactor1967\vcns\work\020303B.BAS"

```
v# = 50

dist# = 1

RANDOMIZE TIMER

CLS

DO

z = INT(RND * 2) + 0

test1# = (v# / 2) - INT(v# / 2)

test2# = (dist# / 2) - INT(dist# / 2)

test3 = (test1# = 0) AND (test2# = 0)

test4 = (test1# = 0) AND (test2# = .5)

test5 = (test1# = .5) AND (test2# = .5)

test6 = (test1# = .5) AND (test2# = 0)

IF test3 = -1 THEN dist# = dist# + 1

IF test4 = -1 THEN dist# = dist# + 2

IF test5 = -1 THEN dist# = dist# + 1

IF test6 = -1 THEN dist# = dist# + 2

r# = v# - dist#

IF z = 0 THEN v# = v# + (v# - r# + 1)

IF z = 1 THEN v# = v# + (v# - r# + 2)

PRINT , z; v#; r#; v# - r#; dist#; v# - sv#; r# - sr#

sv# = v#

sr# = r#

a$ = INKEY$

IF a$ = "s" THEN STOP
```

```
LOOP
```

```
"C:\Users\Reactor1967\vcns\work\020303A.BAS"
REM to calculate your speed# of r#
REM take your dist# without N. Divide it by
REM 4 then multiply it times two.
REM so speed# r# = (dist# / (base * base)) * base
a# = 2
CLS
DO
d# = (a# / 2)
test# = (d# / 2) - INT(d# / 2)
IF test# = .5 THEN d# = d# + 1
s# = a# - d#
PRINT a#; d#; s#; "|"; a# + 1; d# + 1; s#
INPUT a$
IF a$ = "s" THEN STOP
a# = a# + 2
LOOP
```

```
"C:\Users\Reactor1967\vcns\work\020203A.BAS"

v# = 65

r# = 51

CLS

DO

z = INT(RND * 2) + 0

IF z = 0 THEN v# = v# + (v# - r# + 1)

IF z = 1 THEN v# = v# + (v# - r# + 2)

PRINT , , z; v#; r#; v# - r#; v# - sv#; r# - sr#

a$ = INKEY$

INPUT a$

IF a$ = "s" THEN STOP

sv# = v#

sr# = r#

dist# = (v# - r#)

x# = dist#

IF (x# / 2) - INT(x# / 2) = .5 THEN x# = x# - 1

x# = x# / 2

IF (x# / 2) - INT(x# / 2) = .5 THEN x# = x# - 1

r# = r# + (dist# - x#)

dist# = (v# - r#)

LOOP
```

```
"C:\Users\Reactor1967\vcns\work\020103C.BAS"

v# = 50

d1# = 1

d2# = 2

CLS

RANDOMIZE TIMER

DO

z = INT(RND * 2) + 0

IF z = 0 THEN dist# = d1#

IF z = 1 THEN dist# = d2#

IF z = 0 THEN dist# = dist# - 1

IF z = 1 THEN dist# = dist# - 2

dist# = dist# / 2

r# = v# - dist#

REM IF (r# / 2) - INT(r# / 2) = 0 THEN r# = r# - 1

IF z = 0 THEN v# = v# + (v# - r# + 1)

IF z = 1 THEN v# = v# + (v# - r# + 2)

PRINT z; v#; r#; v# - r#; dist#; d1#; d2#; v# - sv#; r# - sr#; v# - sr#

sv# = v#

sr# = r#

a$ = INKEY$

IF a$ = "s" THEN STOP
```

```
d1# = d1# + 2
d2# = d2# + 2
LOOP
```

```
"C:\Users\Reactor1967\vcns\work\020103B.BAS"
v# = 1
r# = 1
CLS
RANDOMIZE TIMER
DO
REM get your dist# + n increase r# by velocity with a chart
REM then code your v# with your d# + n. When decoding look
REM at your dist# + N. Go to the chart and see what your
REM decrease velocity is for r#
z = INT(RND * 2) + 0
IF z = 0 THEN v# = v# + (v# - r# + 1)
IF z = 1 THEN v# = v# + (v# - r# + 2)
PRINT , , z; v#; r#; v# - r#; v# - sv#; r# - sr#

sv# = v#
sr# = r#
dist# = (v# - r#)
```

```
IF (dist# / 2) - INT(dist# / 2) = .5 THEN dist# = dist# - 1
dist# = dist# / 2
dist# = dist# + 1
r# = r# + dist#
IF (r# / 2) - INT(r# / 2) = 0 THEN r# = r# - 1
a$ = INKEY$
REM INPUT a$
IF a$ = "s" THEN STOP
LOOP
```

```
"C:\Users\Reactor1967\vcns\work\020103A.BAS"
v# = 50
dist# = 5
CLS
RANDOMIZE TIMER
DO
z = INT(RND * 2) + 0
test1 = (v# / 2) - INT(v# / 2)
test2 = (dist# / 2) - INT(dist# / 2)
IF test1 = test2 THEN dist# = dist# + 1 ELSE dist# = dist# + 2
r# = v# - dist#
IF z = 0 THEN v# = v# + (v# - r# + 1)
```

```
IF z = 1 THEN v# = v# + (v# - r# + 2)

PRINT , , z; v#; r#; v# - r#; dist#; v# - sv#; r# - sr#

sv# = v#

sr# = r#

a$ = INKEY$

IF a$ = "s" THEN STOP

LOOP
```

"C:\Users\Reactor1967\vcns\work\013103B.BAS"

```
a# = 6

b# = 7

v# = 50

RANDOMIZE TIMER

CLS

DO

z = INT(RND * 2) + 0

dist# = (v# - r#)

test = (dist# / 2) - INT(dist# / 2)

IF test = 0 THEN dist# = a#

IF test = .5 THEN dist# = b#

IF test = 0 THEN dist# = (dist# - 2) / 2

IF test = .5 THEN dist# = (dist# - 1) / 2
```

```
r# = v# - dist#
IF z = 0 THEN v# = v# + (v# - r# + 1)
IF z = 1 THEN v# = v# + (v# - r# + 2)
PRINT z; v#; r#; v# - r#; a#; b#; dist#; v# - sv#; r# - sr#
a$ = INKEY$
IF a$ = "s" THEN STOP
sv# = v#
sr# = r#
a# = a# + 4
b# = b# + 4
LOOP
```

```
"C:\Users\Reactor1967\vcns\work\013103A.BAS"
REM DIST# CAL FOR R#  ALWAYS = .5 WHEN DIVIDED BY 2
C# = 2
D# = 3
CLS
DO
X1# = C#
X1# = X1# - 2
X1# = X1# / 2
```

```
X2# = D#
X2# = X2# - 1
X2# = X2# / 2
PRINT , , C#; X1#, D#; X2#
C# = C# + 4
D# = D# + 4
INPUT A$
IF A$ = "S" THEN STOP
A# = A# + 4
B# = B# + 4
C# = C# + 4
D# = D# + 4
LOOP
```

```
"C:\Users\Reactor1967\vcns\work\013003I.BAS"
v# = 50
r# = 41
dist# = 1
CLS
RANDOMIZE TIMER
DO
z = INT(RND * 2) + 0
```

```
dist# = (v# - r#)
x# = v# - r#
r# = v# - dist#
test1 = (dist# / 2) - INT(dist# / 2) = .5
test2 = (dist# / 2) - INT(dist# / 2) = 0
IF test1 = -1 THEN dist# = ((dist# - 1) / 2) + 1
IF test2 = -1 THEN dist# = dist# / 2
r# = (v# - dist#)
IF (r# / 2) - INT(r# / 2) = 0 THEN r# = r# - 1
IF z = 0 THEN v# = v# + (v# - r# + 1)
IF z = 1 THEN v# = v# + (v# - r# + 2)
PRINT , z; v#; r#; v# - r#; dist#; v# - sv#; r# - sr#
sv# = v#
sr# = r#
a$ = INKEY$
IF a$ = "s" THEN STOP
LOOP
```

```
"C:\Users\Reactor1967\vcns\work\013003H.BAS"
v# = 50
dist# = 1
```

```
CLS
RANDOMIZE TIMER
DO
z = INT(RND * 2) + 0
r# = v# - dist#
IF z = 0 THEN v# = v# + (v# - r# + 1)
IF z = 1 THEN v# = v# + (v# - r# + 2)
REM z = INT(RND * 2) + 0
IF z = 0 THEN r# = r# + (dist# + 1)
IF z = 1 THEN r# = r# + (dist# + 2)
PRINT z; v#; r#; v# - r#; dist#; v# - sv#; r# - sr#
sv# = v#
sr# = r#
a$ = INKEY$
IF a$ = "s" THEN STOP
test1 = (v# / 2) - INT(v# / 2)
test2 = (dist# / 2) - INT(dist# / 2)
IF test1 = test2 THEN dist# = dist# + 1 ELSE dist# = dist# + 2
LOOP
```

"C:\Users\Reactor1967\vcns\work\013003G.BAS"

```
v# = 50

dist# = 5

CLS

RANDOMIZE TIMER

DO

z = INT(RND * 2) + 0

IF z = 0 THEN dist# = dist# + 1

IF z = 1 THEN dist# = dist# + 2

r# = v# - dist#

x1# = v# + (v# - r# + 1)

x2# = v# + (v# - r# + 2)

x1 = (x1# / 2) - INT(x1# / 2)

x2 = (x2# / 2) - INT(x2# / 2)

test1 = (z = 0) AND (x1 = 0)

TEST2 = (z = 0) AND (x2 = 0)

test3 = (z = 1) AND (x1 = .5)

test4 = (z = 1) AND (x2 = .5)

IF test1 = -1 THEN v# = v# + (v# - r# + 1)

IF TEST2 = -1 THEN v# = v# + (v# - r# + 2)

IF test3 = -1 THEN v# = v# + (v# - r# + 1)

IF test4 = -1 THEN v# = v# + (v# - r# + 2)

PRINT , z; v#; r#; v# - r#; dist#; r# - spr#; v# - spv#

a$ = INKEY$

IF a$ = "s" THEN STOP
```

```
spr# = r#

spv# = v#

LOOP
```

"C:\Users\Reactor1967\vcns\work\013003F.BAS"

```
REM goal here was to subtract (v# - r#) - dist# - current N - previous n = speed to subtract from r to get previous r

v# = 1

r# = 1

dist# = 1

z = 0

CLS

RANDOMIZE TIMER

DO

z = INT(RND * 2) + 0

IF z = 0 THEN v# = v# + (v# - r# + 1)

IF z = 1 THEN v# = v# + (v# - r# + 2)

PRINT z; v#; r#; v# - r#; dist#; v# - sv#; r# - sr#

sv# = v#

sr# = r#
```

```
dist# = dist# + 1

r# = v# - dist#

a$ = INKEY$

IF a$ = "s" THEN STOP

LOOP
```

"C:\Users\Reactor1967\vcns\work\013003E.BAS"

```
REM goal here was to subtract (v# - r#) - dist# - current N - previous n = speed to subtract from r to get previous r

v# = 20

r# = 3

z = 0

CLS

RANDOMIZE TIMER

DO

z = INT(RND * 2) + 0

dist# = (v# - r#)

IF (dist# / 2) - INT(dist# / 2) = .5 THEN dist# = dist# - 1

dist# = dist# / 2

dist# = dist# + 2

DO
```

```
spr# = (v# - r#) - dist#

IF z = 0 THEN x# = dist# + 1

IF z = 1 THEN x# = dist# + 2

test1 = (x# - spr# = 0) AND (((v# - dist#) / 2) - INT((v# - dist#) / 2) = .5)

test2 = (x# - spr# = 1) AND (((v# - dist#) / 2) - INT((v# - dist#) / 2) = .5)

IF test1 = -1 THEN EXIT DO

IF test2 = -1 THEN EXIT DO

dist# = dist# - 1

IF dist# = -1 THEN STOP

LOOP

r# = v# - dist#

IF (r# / 2) - INT(r# / 2) = 0 THEN r# = r# - 1

IF z = 0 THEN v# = v# + (v# - r# + 1)

IF z = 1 THEN v# = v# + (v# - r# + 2)

PRINT , , z; v#; r#; v# - r#; v# - spv1#; r# - spr1#

spv1# = v#

spr1# = r#

a$ = INKEY$

IF a$ = "s" THEN STOP

LOOP
```

"C:\Users\Reactor1967\vcns\work\013003D.BAS"

```
REM goal here was to subtract (v# - r#) - dist# - current N - previous n =
speed to subtract from r to get previous r

v# = 20

r# = 3

z = 0

CLS

RANDOMIZE TIMER

DO

IF z = 0 THEN x = 1

IF z = 1 THEN x = 2

z = INT(RND * 2) + 0

IF z = 0 THEN x2 = 1

IF z = 1 THEN x2 = 2

d1# = (v# - r#)

count = (v# - r#)

IF (count / 2) - INT(count / 2) = .5 THEN count = count - 1

count = count / 2

count = count + 1

DO

test1 = (((count * 2) + x2) - (count + x2) - x = ((v# - r#) - count))

IF test1 = -1 THEN EXIT DO

count = count - 1

IF count = 0 THEN EXIT DO

a$ = INKEY$
```

```
IF a$ = "s" THEN STOP
LOOP
IF count = 0 THEN STOP
r# = v# - count
IF (r# / 2) - INT(r# / 2) = 0 THEN r# = r# - 1
dy# = count
IF z = 0 THEN dy# = dy# + 1
IF z = 1 THEN dy# = dy# + 2
IF z = 0 THEN v# = v# + (v# - r# + 1)
IF z = 1 THEN v# = v# + (v# - r# + 2)
PRINT , z; v#; r#; v# - r#; dy#; r# - sr#
a$ = INKEY$
IF a$ = "s" THEN STOP
IF a$ = "d" THEN EXIT DO
sv# = v#
sr# = r#
LOOP
```

```
"C:\Users\Reactor1967\vcns\work\013003C.BAS"
v# = 5
r# = 3
dist# = 2
```

```
CLS

RANDOMIZE TIMER

DO

z = INT(RND * 2) + 0

test1 = (v# / 2) - INT(v# / 2)

IF test1 = .5 THEN test1 = 1

IF z = 0 THEN v# = v# + (v# - r# + 1)

IF z = 1 THEN v# = v# + (v# - r# + 2)

PRINT , , z; v#; r#; v# - r#; v# - sv#; r# - sr#

sr# = r#

sv# = v#

a$ = INKEY$

IF a$ = "s" THEN STOP

IF z = 0 THEN dist# = dist# + 1

IF z = 1 THEN dist# = dist# + 2

r# = r# + (v# - r#) - dist#

IF (r# / 2) - INT(r# / 2) = 0 THEN dist# = dist# + 1

IF (r# / 2) - INT(r# / 2) = 0 THEN r# = r# + (v# - r#) - dist#

LOOP
```

"C:\Users\Reactor1967\vcns\work\013003B.BAS"

```
v# = 6
d# = 10
RANDOMIZE TIMER
CLS
DO
z = INT(RND * 2) + 0
x# = d#
IF z = 0 THEN x# = x# - 2
IF z = 1 THEN x# = x# - 6
x# = x# / 2
r# = v# - x#
IF z = 0 THEN v# = v# + (v# - r# + 2)
IF z = 1 THEN v# = v# + (v# - r# + 6)
IF z = 0 THEN t1# = r# + 2
IF z = 1 THEN t1# = r# + 6
PRINT z; v#; r#; d#; x#; v# - sv#; r# - sr#
a$ = INKEY$
IF a$ = "s" THEN STOP
sv# = v#
sr# = r#
d# = d# + 4
LOOP
```

```
"C:\Users\Reactor1967\vcns\work\013003A.BAS"

v# = 50

dist# = 2

CLS

RANDOMIZE TIMER

DO

z = INT(RND * 2) + 0

r# = v# - dist#

IF z = 0 THEN v# = v# + (v# - r# + 1)

IF z = 1 THEN v# = v# + (v# - r# + 2)

PRINT , z; v#; r#; v# - r#; dist#; r# - spr#; v# - spv#

a$ = INKEY$

IF a$ = "s" THEN STOP

spr# = r#

spv# = v#

test1 = (v# / 2) - INT(v# / 2)

test2 = (dist# / 2) - INT(dist# / 2)

IF test1 = test2 THEN dist# = dist# + 1 ELSE dist# = dist# + 2

LOOP
```

```
"C:\Users\Reactor1967\vcns\work\12803E.BAS"
v# = 50
r# = 45
CLS
RANDOMIZE TIMER
DO
dist# = 1
z = INT(RND * 2) + 0

DO
IF z = 0 THEN x# = dist# + 1
IF z = 1 THEN x# = dist# + 2
sr# = (v# - r#) - dist#
IF x# = sr# THEN EXIT DO
IF x# - sr# = 1 THEN EXIT DO
REM IF x# - sr# = 2 THEN EXIT DO
dist# = dist# + 1
a$ = INKEY$
IF a$ = "s" THEN STOP
LOOP
r# = v# - dist#
REM IF (r# / 2) - INT(r# / 2) = 0 THEN r# = r# - 1
IF z = 0 THEN v# = v# + (v# - r# + 1)
```

```
IF z = 1 THEN v# = v# + (v# - r# + 2)

PRINT , , z; v#; r#; v# - r#; dist#; v# - sv1#; r# - sr1#

sv1# = v#

sr1# = r#

a$ = INKEY$

IF a$ = "s" THEN STOP

LOOP
```

"C:\Users\Reactor1967\vcns\work\12803D.BAS"

```
v# = 50

r# = 45

CLS

RANDOMIZE TIMER

DO

dist# = 1

z = INT(RND * 2) + 0


DO

IF z = 0 THEN x# = dist# + 1

IF z = 1 THEN x# = dist# + 2

sr# = (v# - r#) - dist#

IF x# = sr# THEN EXIT DO
```

```
IF x# - sr# = 1 THEN EXIT DO

REM IF x# - sr# = 2 THEN EXIT DO

dist# = dist# + 1

a$ = INKEY$

IF a$ = "s" THEN STOP

LOOP

r# = v# - dist#

REM IF (r# / 2) - INT(r# / 2) = 0 THEN r# = r# - 1

IF z = 0 THEN v# = v# + (v# - r# + 1)

IF z = 1 THEN v# = v# + (v# - r# + 2)

PRINT , , z; v#; r#; v# - r#; dist#; v# - sv1#; r# - sr1#

sv1# = v#

sr1# = r#

a$ = INKEY$

IF a$ = "s" THEN STOP

LOOP
```

"C:\Users\Reactor1967\vcns\work\012803I.BAS"

```
dist# = 1

CLS

DO

vr1# = (dist# * 2) + 1
```

```
vr2# = (dist# * 2) + 2

sv1# = dist# + 1

sv2# = dist# + 2

PRINT , "0"; vr1#; sv1#; "|"; dist#; "|"; "1"; vr2#; sv2#

a$ = INKEY$

IF a$ = "s" THEN STOP

dist# = dist# + 1

LOOP
```

"C:\Users\Reactor1967\vcns\work\012803H.BAS"

```
v# = 50

r# = 45

spr# = 2

CLS

RANDOMIZE TIMER

DO

Z = INT(RND * 2) + 0

IF Z = 0 THEN v# = v# + (v# - r# + 1)

IF Z = 1 THEN v# = v# + (v# - r# + 2)

PRINT , , Z; v#; r#; v# - r#; spr#

a$ = INKEY$

REM INPUT a$
```

```
IF a$ = "s" THEN STOP
dist# = (v# - r#)
IF (dist# / 2) - INT(dist# / 2) = .5 THEN dist# = dist# - 1
dist# = dist# / 2
IF dist# - spr# >= 2 THEN spr# = spr# + (dist# - spr#)
r# = r# + spr#
LOOP
```

```
"C:\Users\Reactor1967\vcns\work\012803G.BAS"
v# = 50
r# = 45
CLS
RANDOMIZE TIMER
DO
dist# = 1
z = INT(RND * 2) + 0

DO
IF z = 0 THEN x# = dist# + 1
IF z = 1 THEN x# = dist# + 2
```

```
sr# = (v# - r#) - dist#

IF x# = sr# THEN EXIT DO

IF x# - sr# = 1 THEN EXIT DO

REM IF x# - sr# = 2 THEN EXIT DO

dist# = dist# + 1

a$ = INKEY$

IF a$ = "s" THEN STOP

LOOP

r# = v# - dist#

REM IF (r# / 2) - INT(r# / 2) = 0 THEN r# = r# - 1

IF z = 0 THEN v# = v# + (v# - r# + 1)

IF z = 1 THEN v# = v# + (v# - r# + 2)

PRINT , , z; v#; r#; v# - r#; dist#; v# - sv1#; r# - sr1#

sv1# = v#

sr1# = r#

a$ = INKEY$

IF a$ = "s" THEN STOP

LOOP
```

"C:\Users\Reactor1967\vcns\work\012803F.BAS"

```
RANDOMIZE TIMER

REM modulating dist# for coding data into r# and v#
```

```
REM when decoding r# look at (v# - r#) - dist# = N

v# = 7

r# = 3

dist# = 4

z = 0

CLS

DO

REM z1 = INT(RND * 2) + 0

z1 = z2

IF z1 = 0 THEN r# = (v# - dist# - 1)

IF z1 = 1 THEN r# = (v# - dist# - 2)

PRINT , "coding r#"; z1; v#; r#; v# - r#; dist#; v# - sv1#; r# - sr1#

sv1# = v#

sr1# = r#

z2 = INT(RND * 2) + 0

IF z2 = 0 THEN v# = v# + (v# - r# + 1)

IF z2 = 1 THEN v# = v# + (v# - r# + 2)

PRINT , "coding v#"; z2; v#; r#; v# - r#; dist#; v# - sv2#; r# - sr2#

sv2# = v#

sr2# = r#

a$ = INKEY$

IF a$ = "s" THEN STOP

dist# = dist# + 2

LOOP
```

```
"C:\Users\Reactor1967\vcns\work\012803E.BAS"

RANDOMIZE TIMER

v# = 1

dist# = 1

REM equation demonstration program

REM speed of r# = previous(v# - r#) - current dist#

REM speed of v# = (dist# + n)

REM (v# - r#) = (dist# * 2) + n

REM if dist# gets bigger v# - r# gets bigger

REM if dist# gets smaller v# - r# gets smaller

REM using dist you can control most factors of this program

REM it is also possible to code data into r# as well as v# by

REM modulating v# - (dist#+or-n) as well as modulating v# + (dist# + -
n)

REM now try to put all this together into something useful.

z = 0

CLS

DO

z1 = INT(RND * 2) + 0

IF dist# >= 1000 THEN z = 1

IF dist# <= 100 THEN z = 0
```

```
IF z = 0 THEN dist# = dist# + 1

IF z = 1 THEN dist# = dist# - 1

IF z = 0 THEN v# = v# + dist# + 1

IF z = 1 THEN v# = v# + dist# + 2

x# = dist#

IF z1 = 0 THEN x# = (dist# * 2) + 1

IF z1 = 1 THEN x# = (dist# * 2) + 2

r# = v# - x#

PRINT , , z1; v#; r#; v# - r#; dist#; v# - sv#; r# - sr#

sv# = v#

sr# = r#

a$ = INKEY$

IF a$ = "s" THEN STOP

LOOP
```

"C:\Users\Reactor1967\vcns\work\012803D.BAS"

```
RANDOMIZE TIMER

v# = 9

r# = 5

CLS

DO

z = INT(RND * 2) + 0
```

```
dist# = (v# - r#)

IF (dist# / 2) - INT(dist# / 2) = .5 THEN dist# = dist# - 1

dist# = dist# / 2

IF z = 0 THEN dist# = dist# + 1

IF z = 1 THEN dist# = dist# + 2

r# = (v# - dist#)

IF z = 0 THEN v# = v# + (v# - r# + 1)

IF z = 1 THEN v# = v# + (v# - r# + 2)

PRINT , , z; v#; r#; v# - r#; dist#; v# - sv#; r# - sr#

sv# = v#

sr# = r#

a$ = INKEY$

IF a$ = "s" THEN STOP

LOOP
```

"C:\Users\Reactor1967\vcns\work\012803C.BAS"

```
v# = 50

dist# = 2

CLS

RANDOMIZE TIMER

DO

z = INT(RND * 2) + 0
```

```
IF z = 0 THEN dist# = dist# + 2

IF z = 1 THEN dist# = dist# + 4

x# = (dist# / 2) - 2

REM IF z = 0 THEN x# = x# - 2

REM IF z = 1 THEN x# = x# - 4

REM x# = x# / 2

r# = (v# - x#)

IF z = 0 THEN v# = v# + (v# - r# + 2)

IF z = 1 THEN v# = v# + (v# - r# + 4)

PRINT , , z; v#; r#; "vr"; v# - r#; "d"; x#; "sv"; v# - sv#; "sr"; r# - sr#

sv# = v#

sr# = r#

a$ = INKEY$

IF a$ = "s" THEN STOP

LOOP
```

```
"C:\Users\Reactor1967\vcns\work\012803B.BAS"

v# = 50

r# = 46

CLS

RANDOMIZE TIMER
```

```
DO
dist# = (v# - r#)
IF (dist# / 2) - INT(dist# / 2) = .5 THEN dist# = dist# - 1
dist# = dist# / 2
sr# = (v# - r#) - dist#
r# = r# + dist#
z = INT(RND * 2) + 0
IF z = 0 THEN v# = v# + dist# + 2
IF z = 1 THEN v# = v# + dist# + 4
PRINT , , z; v#; r#; v# - r#; dist#; v# - sv2#; r# - sr2#
sv2# = v#
sr2# = r#
a$ = INKEY$
IF a$ = "s" THEN STOP
LOOP
```

```
"C:\Users\Reactor1967\vcns\work\012803A.BAS"
v# = 50
r# = 45
CLS
RANDOMIZE TIMER
```

```
DO
z = INT(RND * 2) + 0
dist# = (v# - r#)
dist# = INT(dist# / 2)
r# = (v# - dist#)
IF z = 0 THEN v# = v# + (v# - r# + 1)
IF z = 1 THEN v# = v# + (v# - r# + 2)
PRINT , , z; v#; r#; v# - r#; dist#; v# - sv#; r# - sr#
sv# = v#
sr# = r#
a$ = INKEY$
IF a$ = "s" THEN STOP
LOOP
```

```
"C:\Users\Reactor1967\vcns\work\012803.BAS"
v# = 50
r# = 45
CLS
RANDOMIZE TIMER
DO
dist# = 1
z = INT(RND * 2) + 0
```

```
DO
IF z = 0 THEN x# = dist# + 1
IF z = 1 THEN x# = dist# + 2
sr# = (v# - r#) - dist#
IF x# = sr# THEN EXIT DO
IF x# - sr# = 1 THEN EXIT DO
REM IF x# - sr# = 2 THEN EXIT DO
dist# = dist# + 1
a$ = INKEY$
IF a$ = "s" THEN STOP
LOOP
r# = v# - dist#
REM IF (r# / 2) - INT(r# / 2) = 0 THEN r# = r# - 1
IF z = 0 THEN v# = v# + (v# - r# + 1)
IF z = 1 THEN v# = v# + (v# - r# + 2)
PRINT , , z; v#; r#; v# - r#; dist#; v# - sv1#; r# - sr1#
sv1# = v#
sr1# = r#
a$ = INKEY$
IF a$ = "s" THEN STOP
LOOP
```

"C:\Users\Reactor1967\vcns\work\012703D.BAS"

```
REM the BEAUTY HERE IS IF YOU KNOW YOUR PREVIOUS (V# - R#) AND

REM YOUR PREVIOUS DISTANCE(work the math to find)

REM USED TO CODE YOU CAN FIND YOUR CURRENT

REM SPEED OF R# TO SUBTRACT your current

REM R# FROM TO GET YOUR PREVIOUS R#

REM PROBLEM HERE IS YOU ALWAYS GOT TO KNOW YOUR PREVIOUS (V# - R#)

REM AFTER CODING

REM if you decode and know your previous n and your previous distance

REM you can do your previous (dist# * 2) + n to find your previous (v# - r#)

REM then just subtract your current dist# before coding from that to find

REM your current speed of r#

CLS

RANDOMIZE TIMER

v# = 7

r# = 7

dist# = 0

DO

test1 = (v# / 2) - INT(v# / 2)

z = INT(RND * 2) + 0

x# = (v# - r#)
```

```
IF z = 0 THEN v# = v# + (v# - r# + 1)

IF z = 1 THEN v# = v# + (v# - r# + 2)

PRINT , z; v#; r#; "vr="; v# - r#; "d="; dist#; v# - sv#; r# - sr#; x#

sv# = v#

sr# = r#

a$ = INKEY$

IF a$ = "s" THEN STOP

IF a$ = "d" THEN EXIT DO

test2 = (v# / 2) - INT(v# / 2)

IF test1 = test2 THEN dist# = dist# + 2

IF test1 <> test2 THEN dist# = dist# + 1

r# = v# - dist#

LOOP

PRINT "Hello"
```

```
"C:\Users\Reactor1967\vcns\work\012703C.BAS"

v# = 50

dist# = 3

r# = v# - dist#

CLS

RANDOMIZE TIMER

DO
```

```
z = INT(RND * 2) + 0

dist# = (v# - r#)

IF z = 0 THEN dist# = dist# + 1

IF z = 1 THEN dist# = dist# + 2

IF z = 0 THEN v# = v# + (v# - r# + 1)

IF z = 1 THEN v# = v# + (v# - r# + 2)

PRINT , , z; v#; r#; v# - r#; v# - sv#; r# - sr#

a$ = INKEY$

IF a$ = "s" THEN STOP

sv# = v#

sr# = r#

IF z = 0 THEN dr# = (dist# - 2)

IF z = 1 THEN dr# = (dist# - 4)

r# = r# + dist#

LOOP
```

```
"C:\Users\Reactor1967\vcns\work\012703B.BAS"

v# = 50

r# = v# - dist#

dist# = 3

CLS

RANDOMIZE TIMER
```

```
DO
z = INT(RND * 2) + 0
IF z = 0 THEN dist# = dist# + 1
IF z = 1 THEN dist# = dist# + 2
IF z = 0 THEN v# = v# + (v# - r# + 1)
IF z = 1 THEN v# = v# + (v# - r# + 2)
PRINT , z; v#; r#; v# - r#; dist#; v# - sv#; r# - sr#
sr# = r#
sv# = v#
r# = (v# - dist#)
REM IF (r# / 2) - INT(r# / 2) = 0 THEN r# = r# - 1
a$ = INKEY$
IF a$ = "s" THEN STOP
LOOP
```

```
"C:\Users\Reactor1967\vcns\work\012703A.BAS"
v# = 1
REM try playing around with this. Change the plus signs to minus signs
REM and vice versa.
r# = 1
dist1# = 0
control = 0
```

```
b1# = 0

CLS

RANDOMIZE TIMER

DO

z = INT(RND * 2) + 0

IF (v# - r#) >= 900 THEN b1# = 1

IF (v# - r#) <= 100 THEN b1# = 0

GOSUB sub1:

REM IF b1# = 0 THEN GOSUB sub1:

REM IF b1# = 1 THEN GOSUB sub2:

r# = r# + ((v# - r#) - dist1#) + 2

v# = v# + dist1#

PRINT z; v#; r#; "vr="; v# - r#; "d="; dist1#; "spv="; v# - spv#; "spr=";
r# - spr#

spv# = (v# - spv#)

spr# = (r# - spr#)

a$ = INKEY$

REM INPUT a$

IF a$ = "s" THEN STOP

IF a$ = "d" THEN EXIT DO

spv# = v#

spr# = r#

LOOP

GOTO decode:
```

```
sub1:

IF z = 0 THEN dist1# = dist1# + 1

IF z = 1 THEN dist1# = dist1# + 2

RETURN

sub2:

IF z = 0 THEN dist1# = dist1# - 1

IF z = 1 THEN dist1# = dist1# - 2

RETURN

decode:

PRINT v#; r#

dist# = v# - r#

dist# = dist# / 2

test1# = (v# / 2) - INT(v# / 2)

x1# = (r# + 1)

x2# = (r# + 2)

x1# = (x1# / 2) - INT(x1# / 2)

x2# = (x2# / 2) - INT(x2# / 2)

PRINT x1#; x2#; test1#
```

```
"C:\Users\Reactor1967\vcns\work\012603B.BAS"

v# = 50

a# = 2
```

```
REM the dist#(v# - r#) = speed# of r# Plus speed# of v# + N

CLS

RANDOMIZE TIMER

DO

z = INT(RND * 2) + 0

DO

a# = a# + 2

IF z = 0 THEN x# = a# - 2

IF z = 1 THEN x# = a# - 4

x# = x# / 2

r# = v# - x#

IF z = 0 THEN x# = v# + (v# - r# + 2)

IF z = 1 THEN x# = v# + (v# - r# + 4)

x# = (x# / 4) - INT(x# / 4)

test1 = ((a# / 4) - INT(a# / 4) = .5) AND (z = 0) AND (x# = 0)

test2 = ((a# / 4) - INT(a# / 4) = 0) AND (z = 1) AND (x# = .5)

IF test1 = -1 THEN EXIT DO

IF test2 = -1 THEN EXIT DO

a$ = INKEY$

IF a$ = "s" THEN STOP

LOOP

IF z = 0 THEN x# = a# - 2

IF z = 1 THEN x# = a# - 4

x# = x# / 2
```

```
r# = v# - x#

IF z = 0 THEN v# = v# + (v# - r# + 2)

IF z = 1 THEN v# = v# + (v# - r# + 4)

PRINT , , z; v#; r#; a#; r# - store#; v# - store2#

store# = r#

store2# = v#

a$ = INKEY$

IF a$ = "s" THEN STOP

LOOP
```

```
"C:\Users\Reactor1967\vcns\work\012603A.BAS"

REM I need one specific r# with each distance

REM there will be two r's# for each N so that

REM v# can be divided by x# to get a fraction value for n>

REM it will look something like this

REM 0 r# = 1 dist# = 3 r# = 2 dist# =5 | 1 r# = 3 dist# = 8 r# = 4 dist# = 10

REM just pick which one you need so that r# is all ways such that

REM v# can be divied by x for a fraction that gives a n value.

REM decoding go down to next line test v for value then test for

REM which r is best in distance such as r# = appropiate value. repeat

REM to keep decoding.
```

```
REM - Lesson learned. Distance / x = N as well so
REM v must = n when divided by x and so does dist#
REM v and x must be oppisite when divided by x
v# = 50
a# = 2
CLS
RANDOMIZE TIMER
DO
z = INT(RND * 2) + 0
DO
a# = a# + 2
x# = a#
IF z = 0 THEN x# = x# - 2
IF z = 1 THEN x# = x# - 4
x# = x# / 2
r# = (v# - x#)
IF z = 0 THEN x# = v# + (v# - r# + 2)
IF z = 1 THEN x# = v# + (v# - r# + 4)
x# = (x# / 4) - INT(x# / 4)
test1 = (z = 0) AND (x# = 0)
test2 = (z = 1) AND (x# = .5)
IF test1 = -1 THEN EXIT DO
IF test2 = -1 THEN EXIT DO
a$ = INKEY$
```

```
IF a$ = "s" THEN STOP

LOOP

IF z = 0 THEN v# = v# + (v# - r# + 2)

IF z = 1 THEN v# = v# + (v# - r# + 4)

REM PRINT , , z; v# / 4; v#; r#; a#; a# / 4

PRINT , , z; v#; r#; a#; v# - r#

a$ = INKEY$

IF a$ = "s" THEN STOP

IF a$ = "d" THEN EXIT DO

LOOP
```

"C:\Users\Reactor1967\vcns\work\012503A.BAS"

```
REM I need one specific r# with each distance

REM there will be two r's# for each N so that

REM v# can be divided by x# to get a fraction value for n>

REM it will look something like this

REM 0 r# = 1 dist# = 3 r# = 2 dist# =5 | 1 r# = 3 dist# = 8 r# = 4 dist# =
10

REM just pick which one you need so that r# is all ways such that

REM v# can be divied by x for a fraction that gives a n value.

REM decoding go down to next line test v for value then test for

REM which r is best in distance such as r# = appropiate value. repeat
```

```
REM to keep decoding.
REM - Lesson learned. Distance / x = N as well so
REM v must = n when divided by x and so does dist#
REM v and x must be oppisite when divided by x
v# = 50
a# = 2
CLS
RANDOMIZE TIMER
DO
z = INT(RND * 2) + 0
DO
a# = a# + 2
test1 = (z = 0) AND ((a# / 4) - INT(a# / 4) = .5)
test2 = (z = 1) AND ((a# / 4) - INT(a# / 4) = 0)
IF test1 = -1 THEN EXIT DO
IF test2 = -1 THEN EXIT DO
a$ = INKEY$
IF a$ = "s" THEN STOP
LOOP
r# = (v# - dist#)
IF z = 0 THEN v# = v# + (v# - r# + 2)
IF z = 1 THEN v# = v# + (v# - r# + 4)
IF z = 0 THEN x# = r# + 2
IF z = 1 THEN x# = r# + 4
```

```
x# = (x# / 4) - INT(x# / 4)

x2# = (v# / 4) - INT(v# / 4)

PRINT , , z; v#; r#; a#; a# / 4

a$ = INKEY$

IF a$ = "s" THEN STOP


LOOP
```

"C:\Users\Reactor1967\vcns\work\012403C.BAS"

```
a# = 6

v# = 2

r# = 0

RANDOMIZE TIMER

CLS

DO

z = INT(RND * 2) + 0

IF z = 0 THEN a# = a# + 2

IF z = 1 THEN a# = a# + 4

IF z = 0 THEN x# = (a# - 2) / 2

IF z = 1 THEN x# = (a# - 4) / 2

r# = v# - x#
```

```
IF z = 0 THEN v# = v# + x# + 2

IF z = 1 THEN v# = v# + x# + 4

PRINT , z; v#; r#; x#; a#

a$ = INKEY$

IF a$ = "s" THEN STOP

LOOP
```

"C:\Users\Reactor1967\vcns\work\012403B.BAS"

```
V# = 50

R# = (V# - A#)

A# = 2

CLS

RANDOMIZE TIMER

DO

Z = INT(RND * 2) + 0

IF Z = 0 THEN V# = V# + (A# + 1)

IF Z = 1 THEN V# = V# + (A# + 2)

IF Z = 0 THEN X1# = (A# * 2) + 1

IF Z = 1 THEN X1# = (A# * 2) + 2

Z = INT(RND * 2) + 0

IF Z = 0 THEN R# = R# + A# + 1

IF Z = 1 THEN R# = R# + A# + 2
```

```
IF Z = 0 THEN X2# = (A# * 2) + 1
IF X = 1 THEN X2# = (A# * 2) + 2
PRINT , V#; V# - X1#; A#; R#; R# - X2#
A$ = INKEY$
IF A$ = "S" THEN STOP
A# = A# + 3
LOOP
```

```
"C:\Users\Reactor1967\vcns\work\012403A.BAS"
r# = 1
v# = 1
CLS
DO
z = INT(RND * 2) + 0
dist# = ABS(v# - r#)
IF z = 0 THEN v# = v# + (dist# + 1)
IF z = 1 THEN v# = v# + (dist# + 2)
z = INT(RND * 2) + 0
IF z = 0 THEN r# = r# + (dist# + 1)
IF z = 1 THEN r# = r# + (dist# + 2)
PRINT ; v#; r#; ABS(v# - r#); dist#
a$ = INKEY$
```

```
IF a$ = "s" THEN STOP
LOOP
```

```
"C:\Users\Reactor1967\vcns\work\012003A.BAS"
v# = 1
r# = 1
speed# = 0
dist# = 0
RANDOMIZE TIMER
CLS
DO
z = INT(RND * 2) + 0
r# = r# + speed#
s1# = v#
IF z = 0 THEN v# = v# + (v# - r# + 1)
IF z = 1 THEN v# = v# + (v# - r# + 2)
PRINT , z; v#; r#; "d ="; dist#; "s ="; speed#; v# - r#
a$ = INKEY$
IF a$ = "s" THEN STOP
t1# = (s1# / 2) - INT(s1# / 2)
t2# = (v# / 2) - INT(v# / 2)
```

```
IF t2# = t1# THEN dist# = dist# + 2

IF t2# <> t1# THEN dist# = dist# + 1

speed# = (v# - r#) - dist#

LOOP
```

"C:\Users\Reactor1967\vcns\work\011803H.BAS"

```
REM DISCOVERY YOU CAN CONTROL THE SPEED OF V# AND R# BY

REM CONTROLING THE SPEED OF R# AND THE DIST# OF (V# - R#)

v# = 5000

r# = 4900

add# = 0

CLS

RANDOMIZE TIMER

DO

z = INT(RND * 2) + 0

dist# = (v# - r#)

IF (dist# / 2) - INT(dist# / 2) = .5 THEN dist# = dist# - 1

dist# = dist# / 2

IF v# - r# >= 1000 THEN add# = 2

IF v# - r# <= 500 THEN add# = 0

dist# = (v# - r#) - (dist# + add#)
```

```
r# = v# - dist#
IF (r# / 2) - INT(r# / 2) = 0 THEN r# = r# - 1
IF z = 0 THEN v# = v# + (v# - r# + 1)
IF z = 1 THEN v# = v# + (v# - r# + 2)
PRINT z; v#; r#; r# - store#; v# - r#; dist#
store# = r#
a$ = INKEY$
REM INPUT a$
IF a$ = "s" THEN STOP
LOOP
```

```
"C:\Users\Reactor1967\vcns\work\011803G.BAS"
v# = 1000
r# = 902
CLS
RANDOMIZE TIMER
x# = 50
DO
dist# = (v# - r#)
dist# = dist# - x#
r# = v# - dist#
z = INT(RND * 2) + 0
```

```
IF z = 0 THEN v# = v# + (v# - r# + 1)

IF z = 1 THEN v# = v# + (v# - r# + 2)

PRINT , , v#; r#; dist#; r# - store#; v# - r#

store# = r#

a$ = INKEY$

INPUT a$

IF a$ = "s" THEN STOP

LOOP
```

```
"C:\Users\Reactor1967\vcns\work\011803F.BAS"

REM SPEED OF R# = (V1# - R1#) - ((V1# - DIST#) = R2#)

v# = 1

r# = 1

dist# = 0

RANDOMIZE TIMER

CLS

DO

store# = r#

z = INT(RND * 2) + 0

r# = v# - dist#

IF z = 0 THEN v# = v# + (v# - r# + 1)

IF z = 1 THEN v# = v# + (v# - r# + 2)
```

```
PRINT , z; v#; r#; r# - store#; dist#; v# - r#
a$ = INKEY$
IF a$ = "s" THEN STOP
dist# = dist# + 2
LOOP




"C:\Users\Reactor1967\vcns\work\011803E.BAS"
v# = 1
r# = 1
CLS
DO
z = INT(RND * 2) + 0
IF z = 0 THEN v# = v# + (v# - r# + 1)
IF z = 1 THEN v# = v# + (v# - r# + 2)
PRINT , , z; v#; r#; v# - r#; dist#
dist# = (v# - r#)
IF z = 0 THEN dist# = dist# - 1
IF z = 1 THEN dist# = dist# - 2
dist# = dist# / 2
dist# = dist# + 1
r# = v# - dist#
a$ = INKEY$
```

```
IF a$ = "s" THEN STOP
LOOP
```

"C:\Users\Reactor1967\vcns\work\011803D.BAS"

```
REM Here R# is much more accurate and does not have to be
REM odd all the time.
v# = 1
d# = 0
CLS
RANDOMIZE TIMER
DO
d# = d# + 1
redo:
x# = d#
z = INT(RND * 2) + 0
IF z = 0 THEN x# = x# + 1
IF z = 1 THEN x# = x# + 2
x2# = v# + x#
test1 = (z = 0) AND ((x2# / 2) - INT(x2# / 2) = 0)
test2 = (z = 1) AND ((x2# / 2) - INT(x2# / 2) = .5)
test3 = (test1 = 0) AND (test2 = 0)
IF test3 = -1 THEN d# = d# + 1
```

```
a$ = INKEY$
IF a$ = "s" THEN STOP
IF test3 = -1 THEN GOTO redo:
v# = x2#
r# = v# - ((d# - 1) * 2)
PRINT , z; v#; d#; r#
LOOP
```

```
"C:\Users\Reactor1967\vcns\work\011803C.BAS"
v# = 2
dist# = 1
CLS
RANDOMIZE TIMER
DO
z = INT(RND * 2) + 0
r# = v# - dist#
IF (r# / 2) - INT(r# / 2) = 0 THEN dist# = dist# + 1
r# = v# - dist#
IF z = 0 THEN v# = v# + (v# - r# + 1)
IF z = 1 THEN v# = v# + (v# - r# + 2)
PRINT , , z; v#; r#; dist#
a$ = INKEY$
```

```
IF a$ = "s" THEN STOP
dist# = dist# + 1
LOOP
```

```
"C:\Users\Reactor1967\vcns\work\011803B.BAS"
v# = 100
dist# = 1
CLS
DO
z = INT(RND * 2) + 0
x# = (dist# * 2)
IF z = 0 THEN x# = x# + 1
IF z = 1 THEN x# = x# + 2
IF z = 0 THEN x# = (x# - 1#) / 2
IF z = 1 THEN x# = (x# - 2#) / 2
r# = v# - x#
IF (r# / 2) - INT(r# / 2) = 0 THEN dist# = dist# + 1
x# = (dist# * 2)
IF z = 0 THEN x# = x# + 1
IF z = 1 THEN x# = x# + 2
IF z = 0 THEN x# = (x# - 1#) / 2
```

```
IF z = 1 THEN x# = (x# - 2#) / 2

r# = v# - x#

IF z = 0 THEN v# = v# + (v# - r# + 1)

IF z = 1 THEN v# = v# + (v# - r# + 2)

PRINT , , z; v#; r#; dist#; v# - r#

a$ = INKEY$

IF a$ = "s" THEN STOP

dist# = dist# + 1

LOOP
```

```
"C:\Users\Reactor1967\vcns\work\011803A.BAS"

v# = 1

r# = 1

dist# = 0

RANDOMIZE TIMER

DO

z = INT(RND * 2) + 0

IF z = 0 THEN v# = v# + (v# - r# + 1)

IF z = 1 THEN v# = v# + (v# - r# + 2)

PRINT , z; v#; r#; dist#; xy#; v# - r#

a$ = INKEY$

IF a$ = "s" THEN STOP
```

```
dist# = dist# + 1
xy# = dist# - 1
r# = v# - xy#
IF (r# / 2) - INT(r# / 2) = 0 THEN xy# = xy# - 1
r# = v# - xy#
a$ = INKEY$
IF a$ = "S" THEN STOP
LOOP
```

"C:\Users\Reactor1967\vcns\work\011703E.BAS"

```
v# = 1
r# = 1
dist# = 0
sy# = 0
xy# = 0
CLS
DO
z = INT(RND * 2) + 0
TEST = (v# / 2) - INT(v# / 2)
test1 = (z = 0) AND (TEST = 0)
test2 = (z = 0) AND (TEST = .5)
test3 = (z = 1) AND (TEST = 0)
```

```
test4 = (z = 1) AND (TEST = .5)

IF test1 = -1 THEN dist# = dist# + 2

IF test2 = -1 THEN dist# = dist# + 1

IF test3 = -1 THEN dist# = dist# + 1

IF test4 = -1 THEN dist# = dist# + 2

xy# = (v# - r#)

IF z = 0 THEN v# = v# + (v# - r#) + 1

IF z = 1 THEN v# = v# + (v# - r#) + 2

PRINT z; v#; r#; "dist# ="; dist#; "xy# ="; xy#; "v# - r# ="; v# - r#

REM PRINT z; v#; r#; xy#

r# = v# - dist#

a$ = INKEY$

REM INPUT a$

IF a$ = "s" THEN STOP

LOOP
```

```
"C:\Users\Reactor1967\vcns\work\011703D.BAS"

v1# = 1

v2# = 1

dist# = 1

CLS
```

```
RANDOMIZE TIMER

DO

z1 = INT(RND * 2) + 0

x# = (dist#)

r1# = (v1# - x#)

IF (r1# / 2) - INT(r1# / 2) = 0 THEN x# = x# + 1

r1# = v1# - x#

IF z1 = 0 THEN v1# = v1# + (v1# - r1# + 1)

IF z1 = 1 THEN v1# = v1# + (v1# - r1# + 2)

z2 = INT(RND * 2) + 0

x# = (dist#)

r2# = (v2# - x#)

IF (r2# / 2) - INT(r2# / 2) = 0 THEN x# = x# + 1

r2# = v2# - x#

IF z2 = 0 THEN v2# = v2# + (v2# - r2# + 1)

IF z2 = 1 THEN v2# = v2# + (v2# - r2# + 2)

PRINT z1; v1#; r1#; v1# - r1#; "|"; dist#; "|"; z2; v2#; r2#; (v2# - r2#)

a$ = INKEY$

IF a$ = "s" THEN STOP

dist# = dist# + 1

LOOP
```

```
"C:\Users\Reactor1967\vcns\work\011703C.BAS"
zy# = 1
v# = 2
CLS
RANDOMIZE TIMER
DO
dist# = zy#
dist# = dist# + 1
r# = v# - dist#
IF (r# / 2) - INT(r# / 2) = 0 THEN dist# = dist# + 1
r# = v# - dist#
z = INT(RND * 2) + 0
IF z = 0 THEN v# = v# + (v# - r# + 1)
IF z = 1 THEN v# = v# + (v# - r# + 2)
PRINT , z; v#; r#; zy#; dist#; v# - r#; r# - store#; v# - store2#
store# = r#
store2# = v#
a$ = INKEY$
IF a$ = "s" THEN STOP
zy# = zy# + 1
LOOP
```

```
"C:\Users\Reactor1967\vcns\work\011703B.BAS"

v# = 1

r# = 1

RANDOMIZE TIMER

zy# = 0

CLS

DO

z = INT(RND * 2) + 0

IF z = 0 THEN v# = v# + (v# - r# + 1)
```

```
"C:\Users\Reactor1967\vcns\work\011303B.BAS"

REM If you know your previous distance the sum of the two

REM Plus X(which can be zero at times) equals then previous (v# - r#)

REM find rules for the x factor.

v# = 1

r# = 1

dist# = 0

RANDOMIZE TIMER

CLS

DO

z = INT(RND * 2) + 0

IF z = 0 THEN v# = v# + (v# - r# + 1)
```

```
IF z = 1 THEN v# = v# + (v# - r# + 2)

PRINT , , z; v#; r#; dist#; v# - r#; r# - store#

store# = r#

dist# = dist# + 1

r# = v# - dist#

IF (r# / 2) - INT(r# / 2) = 0 THEN dist# = dist# + 1

IF (r# / 2) - INT(r# / 2) = 0 THEN r# = v# - dist#

a$ = INKEY$

REM INPUT a$

IF a$ = "s" THEN STOP

LOOP
```

```
"C:\Users\Reactor1967\vcns\work\011303A.BAS"

REM two variables here

REM 1. pick a distance before you code from v#

REM 2. (1) that distance times your base is what you add to r#

REM    with maybe an x variable.

REM 3. know how your distance increments and decrements so you

REM    can use it to decode with it.

REM 4. It might be helpful to have all your r#s divied with the

REM    same decimial value as the remainder so you will have

REM    sysemmettry(missprelled) in your system.
```

```
v# = 10

r# = 9

dist# = 0

CLS

count = 0

DO

z = INT(RND * 2) + 0

dist# = v# - r#

IF z = 0 THEN v# = v# + (v# - r# + 1)

IF z = 1 THEN v# = v# + (v# - r# + 2)

PRINT , z; v#; r#; dist#; count; v# - r#

a$ = INKEY$

REM INPUT a$

IF a$ = "s" THEN STOP

dist# = dist# * 2

r# = r# + dist# - count

count = count + 2

LOOP
```

"C:\Users\Reactor1967\vcns\work\011203B.BAS"

REM (difference in distance) * base = difference in total distance(v# - R#) after coding

```
REM Now, try getting a v and a r and charting

REM it to see if can get an equation for

REM solving for the next r while coding

REM and decoding.

v# = 10

r# = 7

dist# = 3

RANDOMIZE TIMER

CLS

DO

z = INT(RND * 2) + 0

IF z = 0 THEN v# = v# + ((v# - r#) * 3) + 2

IF z = 1 THEN v# = v# + ((v# - r#) * 3) + 4

PRINT z; v#; r#; dist#; dist# - store2#; v# - r#; (v# - r#) - store4#; r# - store#; (r# - store#) - store3#

store4# = v# - r#

store3# = r# - store#

store2# = dist#

store# = r#

DO

dist# = dist# + 1

x# = v# - dist#

IF (x# / 4) - INT(x# / 4) = .75 THEN EXIT DO

a$ = INKEY$
```

```
IF a$ = "s" THEN STOP
LOOP
a$ = INKEY$
IF a$ = "S" THEN STOP
r# = v# - dist#
LOOP
```

"C:\Users\Reactor1967\vcns\work\011203A.BAS"

```
REM ((dist# - n) * 3) = (r2# - r#) if n = 0
REM ((dist# - n) * 3) - 2 = (r2# - r#) if n = 1
REM give all the R's sysmetrry. So that
REM v# can be divided by x and alway know the
REM value of N without finding previous r#.
REM this can be done by having distance
REM increase in relation to v# so that
REM r# / x always as the same decimial value.
REM and im not sure but I think that decimial value
REM realy needs to be in relation to how v# is coded.
v# = 10
dist# = 0
CLS
RANDOMIZE TIMER
```

```
DO

z = INT(RND * 2) + 0

REM IF z = 0 THEN dist# = dist# + 2

REM IF z = 1 THEN dist# = dist# + 4

DO

dist# = dist# + 1

IF ((v# - dist#) / 2) - INT((v# - dist#) / 2) = .5 THEN EXIT DO

a$ = INKEY$

IF a$ = "s" THEN STOP

LOOP

r# = (v# - dist#)

IF z = 0 THEN v# = v# + ((v# - r#) * 3) + 2

IF z = 1 THEN v# = v# + ((v# - r#) * 3) + 4

PRINT , z; v#; r#; v# - r#; dist#; r# - store#

store# = r#

REM PRINT z; (v# / 4); v#; r#; dist#; v# - r#; (r# + 2) / 4; (r# + 4) / 4

a$ = INKEY$

IF a$ = "s" THEN STOP

LOOP
```

"C:\Users\Reactor1967\vcns\work\011103C.BAS"

REM (((v# - r#) / 4) * 2) + 1 (if /4 = 0) + 2 (if / 4 =.5) = (v2# - v1#)

```
v# = 1

r# = 1

dist# = 0

z = 0

CLS

DO

IF z = 0 THEN v# = v# + (v# - r# + 2)

IF z = 1 THEN v# = v# + (v# - r# + 4)

PRINT z; v#; r#; v# - r#; v# - store#

store# = v#

a$ = INKEY$

IF a$ = "s" THEN STOP

z = INT(RND * 2) + 0

IF z = 0 THEN dist# = dist# + 2

IF z = 1 THEN dist# = dist# + 4

IF z = 0 THEN x# = (dist# - 2) / 2

IF z = 1 THEN x# = (dist# - 4) / 2

r# = v# - x#

LOOP
```

"C:\Users\Reactor1967\vcns\work\011103B.BAS"

```
RANDOMIZE TIMER

CLS

v# = 8

r# = 4

z = 0

DO

z = INT(RND * 2) + 0

s1# = v#

x1# = v# + (v# - r# + 1)

x2# = v# + (v# - r# + 2)

test1 = (z = 0) AND ((x1# / 2) - INT(x1# / 2) = 0)

test2 = (z = 1) AND ((x1# / 2) - INT(x1# / 2) = .5)

test3 = (z = 0) AND ((x2# / 2) - INT(x2# / 2) = 0)

test4 = (z = 1) AND ((x2# / 2) - INT(x2# / 2) = .5)

IF test1 = -1 THEN v# = x1#

IF test2 = -1 THEN v# = x1#

IF test3 = -1 THEN v# = x2#

IF test4 = -1 THEN v# = x2#

PRINT , , z; v#; z2; r#; v# - r#; r# - sr#

sr# = r#

REM -------------------------------------------

z = INT(RND * 2) + 0

x1# = v# - 4

x2# = v# - 5
```

```
test1 = (z = 0) AND ((x1# / 2) - INT(x1# / 2) = 0)

test2 = (z = 1) AND ((x1# / 2) - INT(x1# / 2) = .5)

test3 = (z = 0) AND ((x2# / 2) - INT(x2# / 2) = 0)

test4 = (z = 1) AND ((x2# / 2) - INT(x2# / 2) = .5)

IF test1 = -1 THEN r# = x1#

IF test2 = -1 THEN r# = x1#

IF test3 = -1 THEN r# = x2#

IF test4 = -1 THEN r# = x2#

a$ = INKEY$

IF a$ = "s" THEN STOP

z2 = z

LOOP
```

```
"C:\Users\Reactor1967\vcns\work\011103A.BAS"

v# = 10

r# = 9

a# = 4

b# = 5

z2 = 0

CLS

DO

z = INT(RND * 2) + 0
```

```
REM -----------------------------
d2# = (v# - r#)
test1# = v# + (v# - r# + 1)
test2# = v# + (v# - r# + 2)
test3# = (test1# / 2) - INT(test1# / 2)
test4# = (test2# / 2) - INT(test2# / 2)
test5 = (z = 0) AND (test3# = 0)
test6 = (z = 1) AND (test3# = .5)
test7 = (z = 0) AND (test4# = 0)
test8 = (z = 1) AND (test4# = .5)
IF test5 = -1 THEN v# = test1#'
IF test6 = -1 THEN v# = test1#
IF test7 = -1 THEN v# = test2#
IF test8 = -1 THEN v# = test2#
REM -----------------------------
PRINT , , z; v#; z2; r#; v# - r#; d2#; r# - store#
store# = r#
z2 = INT(RND * 2) + 0
REM -----------------------------
x1# = ((v# - a#) / 2) - INT((v# - a#) / 2)
x2# = ((v# - b#) / 2) - INT((v# - b#) / 2)
test1 = (z2 = 0) AND (x1# = 0)
test2 = (z2 = 1) AND (x1# = .5)
test3 = (z2 = 0) AND (x2# = 0)
```

```
test4 = (z2 = 1) AND (x2# = .5)
IF test1 = -1 THEN r# = v# - a#
IF test2 = -1 THEN r# = v# - a#
IF test3 = -1 THEN r# = v# - b#
IF test4 = -1 THEN r# = v# - b#
REM -----------------------------
a$ = INKEY$
REM INPUT a$
IF a$ = "s" THEN STOP
LOOP
```

```
"C:\Users\Reactor1967\vcns\work\011003A.BAS"
v# = 10
d1# = 1
d2# = 2
CLS
RANDOMIZE TIMER
DO
z = INT(RND * 2) + 0
IF z = 0 THEN d1# = d1# + 1
IF z = 0 THEN d2# = d2# + 1
IF z = 1 THEN d1# = d1# + 2
```

```
IF z = 1 THEN d1# = d1# + 2
x1# = d1: x2# = d2#
x1# = x1# - 1 / 2

IF z = 0 THEN v# = v# + (v# - r# + 1)
IF z = 1 THEN v# = v# + (v# - r# + 2)
```

```
"C:\Users\Reactor1967\vcns\work\010803B.BAS"
v# = 100
dist# = 50
CLS
RANDOMIZE TIMER
DO
z = INT(RND * 2) + 0
IF z = 0 THEN x# = (dist# - 2) / 2
IF z = 1 THEN x# = (dist# - 6) / 2
r# = v# - x#
IF z = 0 THEN v# = v# + (v# - r# + 2)
IF z = 1 THEN v# = v# + (v# - r# + 6)
PRINT , , z; v#; r#
store1# = v#
```

```
store2# = r#
a$ = INKEY$
IF a$ = "s" THEN STOP
LOOP
```

```
"C:\Users\Reactor1967\vcns\work\010803A.BAS"
v# = 10
dist# = 2
CLS
RANDOMIZE TIMER
DO
REM try to use the right distances. Thats the key
REM along with the right equations
REM It seems I need a distance data bank
REM pull distances from this in numberical
REM order. If you goto a binary 0 go up 1
REM and use that distance. If you goto a binary
REM 2 go up 2 and use that distance.
REM decode the same way.
REM Look at your binary value and go down
REM 1 in your data bank if you have a binary 0
REM go down 2 in your data bank if you have a
```

```
REM binary 1.

REM work on a better decode test method

REM dividing v# by x and r# + n1 / x and r# + n2 / 2 to test

REM try running r# + n1 and r# + n2 to create a chart to code to the

REM next binary value. Stop when distance is right. Decode using

REM the same chart using it to find the N value of each vector.

z = INT(RND * 2) + 0

IF z = 0 THEN add# = 2

IF z = 1 THEN add# = 6

IF z = 0 THEN dist# = dist# + 2

IF z = 0 THEN dist# = dist# + 6

test = (dist# / 2) - INT(dist# / 2) = .5

IF test = -1 THEN dist# = dist# + 2

x# = dist# - add#

x# = x# / 2

r# = v# - x#

IF z = 0 THEN v# = v# + (v# - r# + 2)

IF z = 1 THEN v# = v# + (v# - r# + 4)

REM PRINT z; (v# / 8); v#; r#; dist#; v# - r#; (r# + 2) / 8; (r# + 6) / 8

PRINT , , z; v#; r#; dist#

a$ = INKEY$

IF a$ = "s" THEN STOP

LOOP
```

```
"C:\Users\Reactor1967\vcns\work\010203B.BAS"

v# = 50

dist# = 1

r# = v# - dist#

CLS

RANDOMIZE TIMER

redo:

DO

z = INT(RND * 2) + 0

IF z = 0 THEN dist# = dist# + 1

IF z = 1 THEN dist# = dist# + 2

r# = v# - dist#

IF z = 0 THEN v# = v# + (v# - r# + 1)

IF z = 1 THEN v# = v# + (v# - r# + 2)

REM PRINT z; v#; z1; r#; v# - r#; dist#; v# - sv#; r# - sr#; v# - sr#

REM LOCATE 12, 10

REM COLOR 2, 0

PRINT z; "+"; v#; r#; v# - r#; "+"; dist#; v# - sv#; r# - sr#; v# - sr#

a$ = INKEY$

IF a$ = "s" THEN STOP

x# = (v# - sv#)
```

```
sv# = v#

sr# = r#

REM test1 = (v# / 2) - INT(v# / 2)

REM test2 = (dist# / 2) - INT(dist# / 2)

REM IF test1 = test2 THEN dist# = dist# + 1 ELSE dist# = dist# + 2

REM z1 = INT(RND * 2) + 0

REM IF z1 = 0 THEN add# = 1

REM IF z1 = 1 THEN add# = 2

REM r# = v# - dist# + add#

REM r# = v# - dist#

LOOP UNTIL dist# >= 1000

REM ------------------------------------

DO

z = INT(RND * 2) + 0

IF z = 0 THEN dist# = dist# - 1

IF z = 1 THEN dist# = dist# - 2

r# = v# - dist#

IF z = 0 THEN v# = v# + (v# - r# + 1)

IF z = 1 THEN v# = v# + (v# - r# + 2)

REM PRINT z; v#; z1; r#; v# - r#; dist#; v# - sv#; r# - sr#; v# - sr#

REM LOCATE 12, 10

REM COLOR 2, 0

PRINT z; "+"; v#; r#; v# - r#; "-"; dist#; v# - sv#; r# - sr#; v# - sr#

a$ = INKEY$
```

```
IF a$ = "s" THEN STOP
sv# = v#
sr# = r#
REM test1 = (v# / 2) - INT(v# / 2)
REM test2 = (dist# / 2) - INT(dist# / 2)
REM IF test1 = test2 THEN dist# = dist# + 1 ELSE dist# = dist# + 2
REM z1 = INT(RND * 2) + 0
REM IF z1 = 0 THEN add# = 1
REM IF z1 = 1 THEN add# = 2
REM r# = v# - dist# + add#
REM r# = v# - dist#
LOOP UNTIL dist# <= 100
GOTO redo:
```

```
"C:\Users\Reactor1967\vcns\work\010203A.BAS"
v# = 1
r# = 1
CLS
RANDOMIZE TIMER
DO
z = INT(RND * 2) + 0
IF z = 0 THEN v# = v# + (v# - r# + 1)
```

```
IF z = 1 THEN v# = v# + (v# - r# + 2)

PRINT , , z; v#; r#; v# - r#; v# - sv#; r# - sr#

IF (v# - r#) >= 100 THEN r# = r# + (v# - sv#)

sv# = v#

sr# = r#

INPUT a$

IF a$ = "s" THEN STOP

LOOP
```

"C:\Users\Reactor1967\vcns\work\010103A.BAS"

```
REM:undate 7 or 8 months later speed of r# needs to be even all the time

REM to make this work.

REM the closer your speed of r# is to your speed of v# and assuming the

REM speed of v# is decent(not too fast or slow)

REM the more stable your program will be.

v# = 1

r# = 1

CLS

RANDOMIZE TIMER

DO

z = INT(RND * 2) + 0

IF z = 0 THEN v# = v# + (v# - r# + 1)
```

```
IF z = 1 THEN v# = v# + (v# - r# + 2)

PRINT , , z; v#; r#; v# - r#; v# - sv#; r# - sr#

IF (v# - r#) >= 100 THEN r# = r# + (v# - sv#)

sv# = v#

sr# = r#

INPUT a$

IF a$ = "s" THEN STOP

LOOP
```

```
"C:\Users\Reactor1967\vcns\work\07300CB.BAS"

d1# = 1

d2# = 2

x1# = 1

x2# = 2

count = 0

CLS

DO

PRINT , , d1#; x1#; d1# - x1#; " "; d2#; x2#; d2# - x2#

INPUT a$

IF a$ = "s" THEN STOP

count = count + 1

IF count = 2 THEN x1# = x1# + 2
```

```
IF count = 2 THEN x2# = x2# + 2

IF count = 2 THEN count = 0

d1# = d1# + 2

d2# = d2# + 2

LOOP
```

"C:\Users\Reactor1967\vcns\work\2123AMRK.BAS"

```
REM this program seems like it might work.

REM notice v# - (r# + increment) is positive when the r# changes

REM notice v# - (v# + increment) is negative when r# does not change.

REM for the above statments it seems true a good part of the time.

v# = 1

r# = 1

CLS

RANDOMIZE TIMER

DO

z = INT(RND * 2) + 0

v# = v# + (v# - r#) + z + 0

PRINT , , (v# - (r# + 50)); v#; r#; v# - r#; v# - sv#; r# - sr#

sv# = v#

sr# = r#

a$ = INKEY$
```

```
IF a$ = "s" THEN STOP

IF v# >= (r# + 50) AND (v# - (r# + 50) < 50) THEN r# = r# + 50

LOOP
```

```
"C:\Users\Reactor1967\vcns\work\VC2003.BAS"

REM when v2# - v1# => 50 then r# = r# + 100

REM try to work out some sort of formula for increasing r2# - r1#

REM with respect to v2# - v1# for coding and decoding.

REM may 29th 2003. It seems you have to create a data base to use this.

REM in the data base store your starting distance, your binary pattern,

REM and your ending distance. As the program codes it uses the
beginning

REM distance and your binary patter to tell when to flip. When it decodes

REM it uses your ending distance and your binary pattern to tell when to

REM flip. You will have to write a program to create this data base. It
will

REM take each distance numerically starting at zero and test a pattern of

REM binary values starting at 0 +plus till the program flips. That
program

REM will store the data base. Keep both of these programs. If this works

REM you can do encryption by having the coding program flip to
different
```

```
REM data base systems as it codes data. Lloyd Burris

REM 061303 I tried the data base and with some success found out that

REM you can create a data base where the speed of v# is the indicator as

REM to what is going on. I discovered the formula

REM (v# - r#) - speed of v# = speed of v# when you decode look at your

REM speed of v# and that would in a database tell you your speed of R
for

REM decode.

CLS

RANDOMIZE TIMER

redo:

great# = 0

less# = 999999

v# = 1

r# = 1

z = 1

g1$ = TIME$

CLS

DO

store# = v#

REM IF (v# + (v# - r#) + 1) - r# = 167 THEN cy = 1 ELSE cy = 0

REM IF (v# + (v# - r#) + 1) - r# = 167 THEN z = INT(RND * 2) + 0

IF z = 0 THEN v# = v# + (v# - r# + 1)

IF z = 1 THEN v# = v# + (v# - r# + 2)
```

```
PRINT , , "+"; z; v#; r#; v# - r#; v# - store#; r# - store2#; great#; less#
IF (v# - r#) > great# THEN great# = v# - r#
IF (v# - r#) < less# THEN less# = v# - r#
REM IF v# - store# = 19 THEN INPUT z$
REM IF z$ = "s" THEN STOP
a$ = INKEY$
REM INPUT a$
IF a$ = "s" THEN t$ = "c:\ps.txt"
IF a$ = "s" THEN fh = 1 ELSE fh = 0
IF fh = 1 THEN CLOSE
IF fh = 1 THEN OPEN t$ FOR OUTPUT AS #1
IF fh = 1 THEN WRITE #1, "+", v#, r#, test
IF fh = 1 THEN CLOSE
IF a$ = "s" THEN STOP
IF a$ = "d" THEN EXIT DO
store2# = r#
IF v# >= (r# + 99) THEN z = 1 ELSE z = 0
IF v# >= (r# + 99) THEN r# = r# + 100
LOOP
g2$ = TIME$
PRINT "Here is the code timer for your program "; g1$; " "; g2$
PRINT "It will take at least this long to dedecode"
INPUT z$
IF z$ = "s" THEN STOP
```

```
test = 0

g3$ = TIME$

DO

z2 = (v# / 2) - INT(v# / 2)

IF z2 = .5 THEN z2 = 1

PRINT , , "-"; z2; v#; r#; v# - r#

IF v# <= 1 THEN EXIT DO

test = (v# / 2) - INT(v# / 2) = .5

REM IF v# - r# = 168 THEN test = 0

a$ = INKEY$

REM INPUT a$

IF a$ = "s" THEN t$ = "c:\ps.txt"

IF a$ = "s" THEN fh = 1 ELSE fh = 0

IF fh = 1 THEN CLOSE

IF fh = 1 THEN OPEN t$ FOR OUTPUT AS #1

IF fh = 1 THEN WRITE #1, "-", v#, r#, test

IF fh = 1 THEN CLOSE

IF a$ = "s" THEN STOP

ghy = ((v# - r#) = 167) OR ((v# - r#) = 168)

d# = v# - r#

IF (d# / 2) - INT(d# / 2) = .5 THEN d# = d# - 1

d# = d# / 2

d# = d# + 1

v# = v# - d#
```

```
IF v# <= 3 THEN test = 0

IF ghy = -1 THEN test = 0

IF test = -1 THEN r# = r# - 100

LOOP

g4$ = TIME$

PRINT "Started coding at "; g1$

PRINT "Stoped codeing at "; g2$

PRINT "Started decoding at "; g3$

PRINT "Stoped decoding at "; g4$
```

```
"C:\Users\Reactor1967\vcns\work\TEMPLATE.BAS"

DECLARE SUB testd (m#, v#, r#, way#)

REM ------------------------TEMPLATE------------------------------

REM DO NOT ALTER THIS PROGRAM IT MIGHT BE USEFUL AS A
TEMPLATE FOR

REM OTHER PROGRAMS!!!!!!!!!!!!!!!!!!!!

REM How to get around my equation wall. Say you want to do base

REM 3 but with base 1 equations meaning that my equations are usually

REM like this v# = v# + ((v# - r#) * base - 1) + N

REM but in base 3 and above its hard to get r# to keep up with v#

REM unless your smarter than I am which is not an impossiblility.

REM so, code like your coding in base 2 but know what to add from
```

```
REM from each N in base to to get base 3 values. Also your using
REM base 3 not base to so n = 1 to 3 instead of n = 1 to 2
REM now I will see if I can do this with out my helper sub.
REM when using a N to tell you when to increment use your lowest N
REM not the highest N
REM 0 to a 0 add 1
REM 0 to a 1 add 1
REM 0 to a 2 add 1
REM 1 to a 0 add 2
REM 1 to a 1 add 2
REM 1 to a 2 add 2
REM 2 to a 0 add 0
REM 2 to a 1 add 0
REM 2 to a 2 add 0
v# = 1
r# = 1
CLS
RANDOMIZE TIMER
REM 0 = 6
REM 1 = 0
REM 2 = 3
DO
z = INT(RND * 2) + 0
z = z + 1
```

```
IF z = 0 THEN z = 1
IF flag = 1 THEN z = 0
way# = v#
IF z = 0 THEN v# = v# + ((v# - r#) * 1) + 1
IF z = 1 THEN v# = v# + ((v# - r#) * 1) + 2
IF z = 2 THEN v# = v# + ((v# - r#) * 1) + 3
way# = v#
DO
m# = (v# / 3) - INT(v# / 3)
m# = m# * 10
m# = INT(m#)
test1 = (z = 0) AND (m# = 6)
test2 = (z = 1) AND (m# = 0)
test3 = (z = 2) AND (m# = 3)
IF test1 = -1 THEN EXIT DO
IF test2 = -1 THEN EXIT DO
IF test3 = -1 THEN EXIT DO
v# = v# + 1
a$ = INKEY$
IF a$ = "s" THEN STOP
LOOP
way# = v# - way#
PRINT , , z; m#; v#; r#; v# - r#; way#
CALL testd(m#, v#, r#, way#)
```

```
a$ = INKEY$
REM INPUT a$
IF a$ = "s" THEN STOP
s1# = r#
IF r# + 94 <= v# THEN r# = r# + 96
IF r# <> s1# THEN flag = 1 ELSE flag = 0
LOOP

SUB testd (m#, v#, r#, way#)
store# = v#
dist# = v# - r#
IF m# = 6 THEN dist# = dist# - 1
END SUB


"C:\Users\Reactor1967\vcns\work\TEMP3.BAS"
CLS
dist# = 100
DO
REM dist# = INT(RND * 1000) + 172
dist# = dist# + 1
d# = 1
DO
```

```
sr# = dist# - d#
sd# = d#
z = INT(RND * 3) + 0
d# = d# * 3
IF z = 0 THEN d# = d# + 1
IF z = 1 THEN d# = d# + 2
IF z = 2 THEN d# = d# + 3
d# = INT(d# / 1.5)
d# = INT(d# / 3)
d# = d# * 3
test = (d# = sr#)
IF test = -1 THEN EXIT DO
d# = sd#
d# = d# + 1
LOOP UNTIL d# = dist#
PRINT , , z; dist#; sd#; sr#; test
a$ = INKEY$
IF a$ = "s" THEN STOP
LOOP
```

```
"C:\Users\Reactor1967\vcns\work\TEMP.BAS"
d# = 1000
```

```
CLS

DO

x# = INT(d# / 2)

DO

s# = d# - x#

IF (s# / 4) - INT(s# / 4) = 0 THEN EXIT DO

x# = x# + 1

a$ = INKEY$

IF a$ = "s" THEN STOP

LOOP

PRINT , d#; x#; s#; (s# / 4) - INT(s# / 4)

d# = d# + 1

s# = d# - x#

PRINT , d#; x#; s#; (s# / 4) - INT(s# / 4)

INPUT a$

IF a$ = "s" THEN STOP

LOOP
```

"C:\Users\Reactor1967\vcns\work\T61103H.BAS"

REM the methods I have been using that is different from coding work

REM at telling me when to increment r but that did not work decoding yet.

```
REM im using the distance at v2 - v1 to tell me when to increment r.
REM im going to try and vary my equations a bit to give me more control
REM over this. L.B.
t$ = "c:\6t1103g.txt"
CLS
RANDOMIZE TIMER
v# = 1
r# = 1
DO
z = INT(RND * 2) + 0
jk# = v#
IF z = 0 THEN v# = v# + (v# - r# + 1)
IF z = 1 THEN v# = v# + (v# - r# + 2)
PRINT , , z; v#; r#; v# - r#; v# - jk#
a$ = INKEY$
IF a$ = "s" THEN STOP
IF a$ = "d" THEN EXIT DO
test = v# >= (r# + 99)
IF test = -1 THEN r# = r# + 100
LOOP
PRINT "-------------------------------"
DO
PRINT , , z; v#; r#; v# - r#; g#
a$ = INKEY$
```

```
INPUT a$
IF a$ = "s" THEN STOP
store# = v#
d# = v# - r#
IF (d# / 2) - INT(d# / 2) = .5 THEN d# = d# - 1
d# = d# / 2
d# = d# + 1
v# = v# - d#
CLOSE #1
OPEN t$ FOR INPUT AS #1
DO
INPUT #1, g#
IF g# = 9999 THEN EXIT DO
test = ((store# - v#) = g#)
IF test = -1 THEN EXIT DO
a$ = INKEY$
IF a$ = "s" THEN STOP
LOOP
IF test = -1 THEN r# = r# - 100
IF v# <= 3 THEN STOP
CLOSE #1
LOOP
```

```
"C:\Users\Reactor1967\vcns\work\T61103G.BAS"
DIM ouch(250)
REM decode
DIM g(250)
count = 1
v# = 1
r# = 1
CLS
RANDOMIZE TIMER
ON ERROR GOTO storeit:
redo:
DO
z = INT(RND * 2) + 0
IF z = 0 THEN v# = v# + (v# - r# + 1)
IF z = 1 THEN v# = v# + (v# - r# + 2)
PRINT , , z; v#; r#; v# - r#; count; v# - store#
a$ = INKEY$
IF a$ = "s" THEN GOTO storeit:
test = v# >= (r# + 99)
IF test = -1 THEN EXIT DO
store# = v#
```

```
LOOP

test = (v# - store#)

FOR dad = 1 TO 250

test2 = (test = g(dad))

IF test2 = -1 THEN EXIT FOR

a$ = INKEY$

IF a$ = "s" THEN GOTO storeit:

NEXT dad

IF test2 = 0 THEN g(count) = v# - store#

IF test2 = 0 THEN count = count + 1

a$ = INKEY$

IF a$ = "s" THEN GOTO storeit:

r# = r# + 100

GOTO redo:

storeit:

t$ = "c:\db.txt"

OPEN t$ FOR OUTPUT AS #1

FOR cc = 1 TO 250

dt = g(cc)

IF dt > 0 THEN WRITE #1, dt

NEXT cc

WRITE #1, v#, r#

CLOSE #1

PRINT "Your done"
```

```
t$ = "c:\6t1103g.txt"

v# = 1

r# = 1

RANDOMIZE TIMER

count = 1

test = 0

CLS

store# = 0

DO

z = INT(RND * 2) + 0

store# = v#

IF z = 0 THEN v# = v# + (v# - r#) + 1

IF z = 1 THEN v# = v# + (v# - r#) + 2

PRINT , , z; v#; r#; v# - r#

IF test = -1 THEN GOSUB ot:

IF count = 251 THEN GOTO wr:

a$ = INKEY$

IF a$ = "s" THEN STOP

test = (v# >= (r# + 99))

IF test = -1 THEN r# = r# + 100

LOOP

STOP

ot:
```

```
test = 0
FOR day = 1 TO 250
g# = ouch(day)
test = (v# - store#) = g#
IF test = -1 THEN EXIT FOR
NEXT day
IF test = 0 THEN ouch(count) = v# - store#
count = count + 1
test = 0
RETURN
wr:
CLOSE #1
OPEN t$ FOR OUTPUT AS #1
FOR day = 1 TO 250
IF ouch(day) > 0 THEN WRITE #1, ouch(day)
NEXT day
WRITE #1, 9999
CLOSE #1
```

"C:\Users\Reactor1967\vcns\work\T61103F.BAS"

```
REM decode
t$ = "c:\db.txt"
```

```
v# = 1
r# = 1
CLS
store# = 0
DO
CLOSE #1
z = INT(RND * 2) + 0
IF z = 0 THEN v# = v# + (v# - r#) + 1
IF z = 1 THEN v# = v# + (v# - r#) + 1
PRINT , , z; v#; r#; v# - r#
a$ = INKEY$
INPUT a$
IF a$ = "s" THEN STOP
IF a$ = "d" THEN GOTO dec:
CLOSE #1
OPEN t$ FOR INPUT AS #1
DO
INPUT #1, g
REM PRINT g
test = (v# - store#) = g
IF test = -1 THEN EXIT DO
LOOP UNTIL g = 999
CLOSE #1
IF test = -1 THEN r# = r# + 100
```

```
store# = v#
LOOP
STOP
dec:
DO
PRINT , , z; v#; r#; v# - r#
a$ = INKEY$
INPUT a$
IF a$ = "s" THEN STOP
store# = v#
d# = (v# - r#)
IF (d# / 2) - INT(d# / 2) = .5 THEN d# = d# - 1
d# = d# / 2
d# = d# + 1
v# = v# - d#
g = v# - store#
CLOSE #1
OPEN t$ FOR INPUT AS #1
DO
INPUT #1, g
test = (v# - store#) = g
IF test = -1 THEN EXIT DO
LOOP UNTIL g = 999
IF test = -1 THEN r# = r# - 100
```

```
CLOSE #1

IF v# <= 3 THEN EXIT DO

LOOP

PRINT , , z; v#; r#; v# - r#




"C:\Users\Reactor1967\vcns\work\T61103E.BAS"

DIM g(250)

count = 1

v# = 1

r# = 1

CLS

RANDOMIZE TIMER

ON ERROR GOTO storeit:

redo:

DO

z = INT(RND * 2) + 0

IF z = 0 THEN v# = v# + (v# - r# + 1)

IF z = 1 THEN v# = v# + (v# - r# + 2)

IF flag = 0 THEN PRINT , , z; v#; r#; v# - r#

IF flag = 1 THEN PRINT , , z; v#; r#; v# - r#; count; v# - store#

a$ = INKEY$

IF a$ = "s" THEN GOTO storeit:
```

```
test = v# >= (r# + 99)
IF test = -1 THEN flag = 1 ELSE flag = 0
IF test = -1 THEN EXIT DO
store# = v#
LOOP
test = (v# - store#)
FOR dad = 1 TO 250
test2 = (test = g(dad))
IF test2 = -1 THEN EXIT FOR
a$ = INKEY$
IF a$ = "s" THEN GOTO storeit:
NEXT dad
IF test2 = 0 THEN g(count) = v# - store#
IF test2 = 0 THEN count = count + 1
a$ = INKEY$
IF a$ = "s" THEN GOTO storeit:
r# = r# + 100
GOTO redo:
storeit:
t$ = "c:\db.txt"
OPEN t$ FOR OUTPUT AS #1
FOR cc = 1 TO 250
dt = g(cc)
IF dt > 0 THEN WRITE #1, dt
```

```
NEXT cc
WRITE #1, v#, r#
CLOSE #1
PRINT "Your done"
STOP
IF z$ = "s" THEN STOP
```

"C:\Users\Reactor1967\vcns\work\T61103.BAS"

```
t1$ = "c:\db.txt"
t2$ = "c:\db2.txt"
CLS
```

"C:\Users\Reactor1967\vcns\work\T060203B.BAS"

```
REM CLOSE #1
REM OPEN t$ FOR INPUT AS #1
allover:
CLS
t$ = "c:\db.txt"
REM INPUT #1, lb#, realdeal$, crank#
v# = 1
```

```
r# = 1

RANDOMIZE TIMER

CLS

start# = (v# - r#)

bin$ = ""

ghz = 0

DO

z = INT(RND * 2) + 0

IF z = 0 THEN bin$ = bin$ + "0"

IF z = 1 THEN bin$ = bin$ + "1"

IF z = 0 THEN v# = v# + ABS(v# - r#) + 1

IF z = 1 THEN v# = v# + ABS(v# - r#) + 2

PRINT , z; v#; r#; v# - r#

a$ = INKEY$

IF v# - r# >= 200 THEN PRINT sh1$, sh2$

IF v# - r# >= 200 THEN INPUT a$

IF a$ = "s" THEN STOP

IF a$ = "a" THEN GOTO allover:

CLOSE #1

OPEN t$ FOR INPUT AS #1

DO

INPUT #1, z1, g$, z2

sh1$ = bin$: sh2$ = g$

meredith = (z1 = start#) AND (bin$ = g$) AND (z2 = (v# - r#))
```

```
IF meredith = -1 THEN EXIT DO

test = (z1 = 999) OR (bin$ = "999") OR (z2 = 999)

IF test = -1 THEN EXIT DO

LOOP

IF meredith = -1 THEN r# = r# + 100

IF meredith = -1 THEN bin$ = ""

IF meredith = -1 THEN start# = (v# - r#)

CLOSE #1

LOOP
```

```
"C:\Users\Reactor1967\vcns\work\T060103.BAS"

REM this program creates a database that vcns can use to function with.

REM the purpose of this data base is to find stable sequences of patterns

REM of numbers that can be run and used to represent 1's and 0's.

REM sequences may be needed to be used as switches to switch from
different

REM patterns. Anyway if a specific pattern can be establish that can
switch

REM back and forth from each other and predictable enough to reconize
thus

REM decode with then it might be possible to code binary data with it.

DECLARE SUB bn (count!, bin$)

CLS
```

```
t$ = "c:\db.txt"
OPEN t$ FOR OUTPUT AS #1
v# = 999
r# = 999
store# = v#
count = 0
LB# = v# - r#
DO
CALL bn(count, bin$)
binstore = count
g = LEN(bin$)
flag = 0
realdeal$ = ""
DO
PRINT , , v#; r#; v# - r#; z
IF g > 0 THEN z = VAL(MID$(bin$, g, 1)) ELSE z = 0
IF z = 0 THEN flag = 1
IF z = 0 THEN realdeal$ = realdeal$ + "0"
IF z = 1 THEN realdeal$ = realdeal$ + "1"
IF z = 0 THEN v# = v# + (v# - r#) + 1
IF z = 1 THEN v# = v# + (v# - r#) + 2
g = g - 1
test = (v# >= (r# + 99))
IF test = -1 THEN EXIT DO
```

```
a$ = INKEY$
IF a$ = "s" THEN STOP
LOOP
PRINT , , v#; r#; v# - r#; z; v# - (r# - 100)
PRINT , , LB#; bin$; (v# - r#); binstore; realdeal$
a$ = INKEY$
REM INPUT a$
IF a$ = "s" THEN STOP
crank# = v# - r#
WRITE #1, (100 + LB#), LB#, realdeal$, crank#, v# - (r# + 100)
count = count + 1
IF flag = 0 THEN count = 0
IF flag = 0 THEN store# = store# + 1
PRINT "                    "
v# = store#
LB# = v# - r#
IF v# - r# >= 99 THEN WRITE #1, 999, "999", 999
IF v# - r# >= 99 THEN CLOSE #1
IF v# - r# >= 99 THEN STOP
LOOP


SUB bn (count, bin$)
IF count < 0 THEN STOP
IF count > 9999999 THEN STOP
```

```
store = count
d# = 1
count2 = 1
DO
IF d# >= count THEN EXIT DO
d# = d# * 2
count2 = count2 + 1
LOOP
IF d# > count THEN count2 = count2 - 1
count2 = count2 - 1
bin$ = ""
DO
a# = (2 ^ count2)
IF a# <= count THEN bin$ = bin$ + "1"
IF a# > count THEN bin$ = bin$ + "0"
IF a# <= count THEN count = count - a#
count2 = count2 - 1
IF count2 < 0 THEN EXIT DO
LOOP
count = store
END SUB
```

```
"C:\Users\Reactor1967\vcns\work\SYMETERY.BAS"

REM this program is helpful for finding what value

REM to use to test your vector numbers

REM add a n value to your r# then enter that as your r value

REM the possible values of a# to use for testing are

REM when your v# and (r# + N) value have the same

REM fractional value.

INPUT "input your r + n value"; r#

INPUT "input your v value"; v#

a# = 1

CLS

DO

PRINT , a#, v# / a#; r# / a#

INPUT a$

IF a$ = "s" THEN STOP

a# = a# + 1

LOOP
```

```
"C:\Users\Reactor1967\vcns\work\SPEEDR.BAS"

v# = 1000

r# = 902

CLS
```

```
RANDOMIZE TIMER

x# = 50

DO

dist# = (v# - r#)

dist# = dist# - x#

r# = v# - dist#

z = INT(RND * 2) + 0

IF z = 0 THEN v# = v# + (v# - r# + 1)

IF z = 1 THEN v# = v# + (v# - r# + 2)

PRINT , , v#; r#; dist#; r# - store#; v# - r#

store# = r#

a$ = INKEY$

INPUT a$

IF a$ = "s" THEN STOP

LOOP
```

```
"C:\Users\Reactor1967\vcns\work\REPEAT2.BAS"

REM Demonstration of the Vector Coordinate Numerical System.

v# = 1

r# = 1

v2# = 1

r2# = 1
```

```
hgj# = 1

CLS

by# = 0

RANDOMIZE TIMER

z = 0

TIMER ON

a1$ = TIME$

DO

REM z = INT(RND * 2) + 0

IF r# < 0 THEN r# = 1

IF r# > v# THEN r# = r# - 100

IF z = 0 THEN v# = v# + (v# - r# + 1)

IF z = 1 THEN v# = v# + (v# - r# + 2)

by# = by# + 1

PRINT , , by#; z; v#; r#; v# - r#; v# - store1#

store1# = v#

a$ = INKEY$

REM INPUT a$

IF a$ = "s" THEN STOP

IF a$ = "d" THEN EXIT DO

sr1# = r#

t# = (v# / 100) - INT(v# / 100)

r# = (v# / 100) - t#

r# = (r# * 100) + 1
```

```
sr2# = r#

IF sr1# = sr2# THEN z = 0

IF sr1# <> sr2# THEN z = 1

LOOP

b$ = TIME$

PRINT "Encode Complete! Number of bits encoded is "; by#; " Hit Enter
to begin decode."

PRINT "starting coding at "; a1$

PRINT "stoped coding at "; b$

INPUT z$

IF z$ = "s" THEN SYSTEM

sby# = by#

REM ------------------------------------------------

REM STOP

c$ = TIME$

DO

test = (v# / 2) - INT(v# / 2)

IF test = 0 THEN z = 0

IF test = .5 THEN z = 1

PRINT , , by#; z; v#; r#; v# - r#

by# = by# - 1

a$ = INKEY$

REM INPUT a$

IF a$ = "s" THEN STOP
```

```
dist# = v# - r#

IF (dist# / 2) - INT(dist# / 2) = .5 THEN dist# = dist# - 1

dist# = dist# / 2

dist# = dist# + 1

v# = v# - dist#

IF z = 1 THEN r# = r# - 100

IF r# <= 0 THEN r# = 1

IF v# <= 1 THEN d$ = TIME$

IF v# <= 1 THEN PRINT , , by#; z; v#; r#; v# - r#

IF v# <= 1 THEN PRINT "Decode Complete!"

IF v# <= 1 THEN PRINT "started coding at "; a1$; " stoped coding at ";
b$

IF v# <= 1 THEN PRINT "started decoding at "; c$; " stoped decoding at
"; d$

IF v# <= 1 THEN PRINT sby#; " Bits decoded."

IF v# <= 1 THEN INPUT z$

IF v# <= 1 THEN SYSTEM

LOOP
```

```
"C:\Users\Reactor1967\vcns\work\REPEAT.BAS"

CLS

RANDOMIZE TIMER

redo:
```

```
v# = 1
r# = 1
z = 1
CLS
DO
store# = v#
REM IF (v# + (v# - r#) + 1) - r# = 167 THEN cy = 1 ELSE cy = 0
REM IF (v# + (v# - r#) + 1) - r# = 167 THEN z = INT(RND * 2) + 0
IF z = 0 THEN v# = v# + (v# - r# + 1)
IF z = 1 THEN v# = v# + (v# - r# + 2)
PRINT , , "+"; z; v#; r#; v# - r#; v# - store#; r# - store2#
REM IF v# - store# = 19 THEN INPUT z$
REM IF z$ = "s" THEN STOP
a$ = INKEY$
REM INPUT a$
IF a$ = "s" THEN t$ = "c:\ps.txt"
IF a$ = "s" THEN fh = 1 ELSE fh = 0
IF fh = 1 THEN CLOSE
IF fh = 1 THEN OPEN t$ FOR OUTPUT AS #1
IF fh = 1 THEN WRITE #1, "+", v#, r#, test
IF fh = 1 THEN CLOSE
IF a$ = "s" THEN STOP
IF a$ = "d" THEN EXIT DO
store2# = r#
```

```
IF v# >= (r# + 99) THEN z = 1 ELSE z = 0

IF v# >= (r# + 99) THEN r# = r# + 100

LOOP

test = 0

DO

z2 = (v# / 2) - INT(v# / 2)

IF z2 = .5 THEN z2 = 1

PRINT , , "-"; z2; v#; r#; v# - r#

IF v# <= 1 THEN EXIT DO

test = (v# / 2) - INT(v# / 2) = .5

REM IF v# - r# = 168 THEN test = 0

a$ = INKEY$

REM INPUT a$

IF a$ = "s" THEN t$ = "c:\ps.txt"

IF a$ = "s" THEN fh = 1 ELSE fh = 0

IF fh = 1 THEN CLOSE

IF fh = 1 THEN OPEN t$ FOR OUTPUT AS #1

IF fh = 1 THEN WRITE #1, "-", v#, r#, test

IF fh = 1 THEN CLOSE

IF a$ = "s" THEN STOP

ghy = ((v# - r#) = 167) OR ((v# - r#) = 168)

d# = v# - r#

IF (d# / 2) - INT(d# / 2) = .5 THEN d# = d# - 1

d# = d# / 2
```

```
d# = d# + 1
v# = v# - d#
IF v# <= 3 THEN test = 0
IF ghy = -1 THEN test = 0
IF test = -1 THEN r# = r# - 100
LOOP
REM STOP
GOTO redo:
```

```
"C:\Users\Reactor1967\vcns\work\RATECHNG.BAS"
REM this program is used for attempting to calculate rate of change
REM for v# - r# while coding up in value. The goal here is to use this for
REM binary 0 to a binary 0 rate of change (v - r) = x1
REM binary 0 to a binary 1 rate of change (v - r) = x2
REM binary 1 to a binary 0 rate of change (v - r) = x3
REM binary 1 to a binary 1 rate of change (v - r) = x4
REM If I can get a specific rate of change for
REM v# - r# for (x1 to x4) then it might be possible
REM to get a decode because I will know how much to
REM subtract from (v - r) to subtract from my previous v
REM to get my previous r and repeat that process.
redo:
```

```
CLS
REM INPUT "Enter starting (v# - r#)"; lb1
REM INPUT "Enter desired rate of change"; rate
d# = 2
d2# = 3
count = 0
DO
REM PRINT lb1; rate
REM PRINT d# - lb1; d2# - lb1
PRINT , , d#; count; x#; d2#; count2; x2#
REM IF d# >= lb1 THEN INPUT a$
REM IF a$ = "s" THEN STOP
REM test1 = ((d# - lb1) >= rate) OR ((d2# - lb1) >= rate)
REM IF test1 = -1 THEN EXIT DO
count = 0
count2 = 0
DO
count = count + 1
d# = d# + 2
x# = d# - 2
x# = x# / 2
IF (x# / 2) - INT(x# / 2) = 0 THEN EXIT DO
a$ = INKEY$
IF a$ = "s" THEN STOP
```

```
LOOP

a$ = INKEY$

IF a$ = "s" THEN STOP

DO

count2 = count2 + 1

d2# = d2# + 2

x2# = d2# - 1

x2# = x2# / 2

IF (x2# / 2) - INT(x2# / 2) = .5 THEN EXIT DO

a$ = INKEY$

IF a$ = "s" THEN STOP

LOOP

a$ = INKEY$

IF a$ = "s" THEN STOP

LOOP

PRINT "----------------------------------------------------"

PRINT lb1; rate

PRINT d# - lb1; d2# - lb1

PRINT , , d#; count; x#; d2#; count2; x2#

INPUT a$

IF a$ = "s" THEN STOP

GOTO redo:
```

```
"C:\Users\Reactor1967\vcns\work\RATECHAN.GE"
REM this program is used for attempting to calculate rate of change
REM for v# - r#. The goal here is to use this for
REM binary 0 to a binary 0 rate of change = x1
REM binary 0 to a binary 1 rate of change = x2
REM binary 1 to a binary 0 rate of change = x3
REM binary 1 to a binary 1 rate of change = x4
REM If I can get a specific rate of change for
REM v# - r# for (x1 to x4) then it might be possible
REM to get a decode
redo:
CLS
INPUT "Enter starting (v# - r#)"; lb1
INPUT "Enter desired rate of change"; rate
d# = 2
d2# = 3
count = 0
DO
PRINT , , d#; count; x#; d2#; count2; x2#
REM INPUT a$
REM IF a$ = "s" THEN STOP
test1 = ((d# - lb1) >= rate) OR ((d2# - lb1) >= rate)
IF test1 = -1 THEN EXIT DO
```

```
count = 0

count2 = 0

DO

count = count + 1

d# = d# + 2

x# = d# - 2

x# = x# / 2

IF (x# / 2) - INT(x# / 2) = .5 THEN EXIT DO

a$ = INKEY$

IF a$ = "s" THEN STOP

LOOP

a$ = INKEY$

IF a$ = "s" THEN STOP

DO

count2 = count2 + 1

d2# = d2# + 2

x2# = d2# - 1

x2# = x2# / 2

IF (x2# / 2) - INT(x2# / 2) = 0 THEN EXIT DO

a$ = INKEY$

IF a$ = "s" THEN STOP

LOOP

a$ = INKEY$

IF a$ = "s" THEN STOP
```

```
LOOP
PRINT "------------------------------------------------------"
PRINT lb1; rate
PRINT d# - lb1; d2# - lb1
PRINT , , d#; count; x#; d2#; count2; x2#
INPUT a$
IF a$ = "s" THEN STOP
GOTO redo:


"C:\Users\Reactor1967\vcns\work\100104C.BAS"
REM This did not work because m# was off. Still playing with key files.
a$ = "c:\rfile.txt"
OPEN a$ FOR OUTPUT AS #1
CLOSE #1
c# = 100
RANDOMIZE TIMER
DO
DO
r# = INT(RND * c#) + 0
test = ((c# - r#) / 3) - INT((c# - r#) / 3) = 0
IF test = -1 THEN EXIT DO
d$ = INKEY$
IF d$ = "s" THEN STOP
LOOP
```

```
OPEN a$ FOR APPEND AS #1

WRITE #1, c#, r#

CLOSE #1

PRINT , , c#; r#

d$ = INKEY$

IF d$ = "s" THEN STOP

c# = c# + 1

LOOP UNTIL c# >= 10000

INPUT "Ready to code."; z$

v# = 100

CLOSE #1

DO

OPEN a$ FOR INPUT AS #1

DO

INPUT #1, a#, b#

IF a# = v# THEN EXIT DO

d$ = INKEY$

IF d$ = "s" THEN STOP

LOOP UNTIL EOF(1)

CLOSE #1

r# = b#

z = INT(RND * 3) + 0

IF z = 0 THEN v# = v# + (v# - r#) + 1

IF z = 1 THEN v# = v# + (v# - r#) + 2
```

```
IF z = 2 THEN v# = v# + (v# - r#) + 3

m# = (v# / 3) - INT(v# / 3): m# = m# * 100: m# = INT(m#)

PRINT , , z; m#; v#; r#; v# - r#

d$ = INKEY$

IF d$ = "s" THEN STOP

LOOP UNTIL v# > 10000
```

"C:\Users\Reactor1967\vcns\work\100104B.BAS"

```
REM This program dimistrates what I was trying to do but it does not work.

v# = 1

r# = 1

CLS

RANDOMIZE TIMER

DO

z = INT(RND * 2) + 0

test = (v# + ((v# - r#) * 2) + z + 1) - r# >= 100

IF test = -1 THEN z = 2

IF z = 0 THEN v# = v# + ((v# - r#) * 2) + 1

IF z = 1 THEN v# = v# + ((v# - r#) * 2) + 2

IF z = 2 THEN v# = v# + ((v# - r#) * 2) + 3

PRINT , , z; v#; r#; v# - r#
```

```
a$ = INKEY$
INPUT a$
IF a$ = "s" THEN STOP
IF z = 2 THEN r# = r# + 100
LOOP
```

"C:\Users\Reactor1967\vcns\work\100104A.BAS"

```
REM What im playing with here is a method to reduce any number and decode
REM it back to its orginal number. The problem here is there has to be
REM a way to know when to stop when decoding a reduced number.
REM This works just great if you know what
REM range you last v# lays in at the end of decoding.
REM its also possible instead of using a program to reduce just have a
REM function built into the coding system that a specific N means
REM reduce r# a specific amount. The only thing here is that
REM ((v# - r#) * base - 1) can not be larger than the amount reducing by.
REM what you would be reducing would be the distance between v# & r# not
REM necessarly r# itself. This would be done by change v# or r# or both.
v# = 4
r# = 1
CLS
```

```
RANDOMIZE TIMER

redo:

DO

z = INT(RND * 3) + 0

IF z = 0 THEN v# = v# + ((v# - r#) * 2) + 1

IF z = 1 THEN v# = v# + ((v# - r#) * 2) + 2

IF z = 2 THEN v# = v# + ((v# - r#) * 2) + 3

m# = ((v# - r#) / 3) - INT((v# - r#) / 3): m# = m# * 10: m# = INT(m#)

PRINT , , z; m#; v#; "+"; r#; v# - r#

a$ = INKEY$

REM INPUT a$

IF a$ = "s" THEN STOP

LOOP UNTIL v# > 10000

r# = 50000

count = 50000

IF v# > 50000 THEN STOP

DO

REM PRINT count

IF count <= (r# - 9) THEN r# = r# - 10

count = count - 1

LOOP UNTIL count <= v#

z = 1

DO

IF v# - (r# - v#) - z - 1 < 1 THEN EXIT DO
```

```
IF z = 0 THEN v# = v# - (r# - v#) - 1

IF z = 1 THEN v# = v# - (r# - v#) - 2

PRINT , , z; v#; "-"; r#; v# - r#

a$ = INKEY$

REM INPUT a$

IF a$ = "s" THEN STOP

test = v# <= (r# - 9)

IF test = -1 THEN r# = r# - 10

IF test = -1 THEN z = 1 ELSE z = 0

LOOP

r# = 1

GOTO redo:
```

"C:\Users\Reactor1967\vcns\work\012504C.BAS"

```
REM So use base 3 for v1# that at least will tell you when to exit that

REM coding process. On v2# some relationship needs established beteen

REM v2# and v1# so that when done decoding v1# you will know when
to switch

REM back to decoding v2#. Looking at this program has some examples
when

REM run and looked at.

REM Problem here is that you can always tell when to look at N for base
2
```

```
REM or look at n for base 3. Thats not easy.

v1# = 1

v2# = 1

r# = 1

CLS

RANDOMIZE TIMER

redo:

z = 2

GOTO iraq:

REM When decoding here v2# = v1# = r# is the out condition.

DO

z = INT(RND * 2) + 0

REM test = ((v1# + ((v1# - r#) * 2) + 3) - r#) >= 99 Don,t do this at end.

REM IF test = -1 THEN z = 2 don,t do this at end

iraq:

IF z = 0 THEN v1# = v1# + ((v1# - r#) * 2) + 1

IF z = 1 THEN v1# = v1# + ((v1# - r#) * 2) + 2

IF z = 2 THEN v1# = v1# + ((v1# - r#) * 2) + 3

m# = (v1# / 3) - INT(v1# / 3)

m# = m# * 10

m# = INT(m#)

PRINT , , "+"; "|"; z; "|"; m#; "|"; v1#; "|"; v2#; "|"; r#; "|"

a$ = INKEY$

REM INPUT a$
```

```
IF a$ = "s" THEN STOP

REM LOOP UNTIL z = 2

LOOP UNTIL v1# - r# >= 99

PRINT "-----------------------------------------------"

REM here v2# begin with a relationship to v1# so when decoding v2#

REM we can look at v1# and tell when to stop decoding v2# and begin

REM coding v1#. v1# takes care of itself in the decoding process.

DO

z = 0

test = (r# + 2) <= v2#

IF test = -1 THEN r# = r# + 3

IF test = -1 THEN z = 1 ELSE z = 0

IF z = 2 THEN sv# = v#

IF z = 0 THEN v2# = v2# + ((v2# - r#) * 1) + 1

IF z = 1 THEN v2# = v2# + ((v2# - r#) * 1) + 2

m# = (v2# / 4) - INT(v2# / 4)

m# = m# * 10

m# = INT(m#)

PRINT , , "-"; "|"; z; "|"; m#; "|"; v1#; "|"; v2#; "|"; r#; "|"

a$ = INKEY$

REM INPUT a$

IF a$ = "s" THEN STOP

LOOP UNTIL v2# >= v1#

PRINT "-----------------------------------------------"
```

```
GOTO redo:

"C:\Users\Reactor1967\vcns\work\012504B.BAS"
v1# = 1
v2# = 1
r# = 1
CLS
RANDOMIZE TIMER
redo:
DO
z = INT(RND * 2) + 0
test = ((v1# + ((v1# - r#) * 2) + 3) - r#) >= 99
IF test = -1 THEN z = 2
IF z = 0 THEN v1# = v1# + ((v1# - r#) * 2) + 1
IF z = 1 THEN v1# = v1# + ((v1# - r#) * 2) + 2
IF z = 2 THEN v1# = v1# + ((v1# - r#) * 2) + 3
m# = (v1# / 3) - INT(v1# / 3)
m# = m# * 10
m# = INT(m#)
PRINT , , z; m#; v1#; v2#; r#
```

```
a$ = INKEY$
REM INPUT a$
IF a$ = "s" THEN STOP
LOOP UNTIL z = 2
DO
z = 0
DO
test = (r# + 3) <= v2#
IF test = -1 THEN z = 1 ELSE z = 0
IF test = -1 THEN r# = r# + 3
a$ = INKEY$: IF a$ = "s" THEN STOP
LOOP UNTIL test = 0
test = v2# + ((v2# - r#) * 2) + 3 > v1#
IF test = -1 THEN z = 2
IF z = 0 THEN v2# = v2# + ((v2# - r#) * 2) + 1
IF z = 1 THEN v2# = v2# + ((v2# - r#) * 2) + 2
IF z = 2 THEN v2# = v2# + ((v2# - r#) * 2) + 3
m# = (v1# / 3) - INT(v1# / 3)
m# = m# * 10
m# = INT(m#)
PRINT , , z; m#; v1#; v2#; r#
a$ = INKEY$
REM INPUT a$
IF a$ = "s" THEN STOP
```

```
LOOP UNTIL z = 2

GOTO redo:

"C:\Users\Reactor1967\vcns\work\012504A.BAS"

REM trying to make this predictable. Look at count it follows a pattern.

REM when coding follow a pattern until two is needed then insert that

REM specificly into the followed pattern and a decode process might be

REM able to develop here. This would be useful in trying to have a
program

REM that can catch r# up to v# and be decodalbe with.

DECLARE SUB catch (v#, r#)

v# = 1

r# = 1

z = 0

CLS

RANDOMIZE TIMER

count = 0

DO

count = count + 1

sv# = v#

IF z = 0 THEN v# = v# + ((v# - r#) * 2) + 1
```

```
IF z = 1 THEN v# = v# + ((v# - r#) * 2) + 2

IF z = 2 THEN v# = v# + ((v# - r#) * 2) + 3

m# = (v# / 3) - INT(v# / 3)

m# = m# * 10

m# = INT(m#)

PRINT , , z; m#; v#; r#; v# - r#; "sr# = "; r# - sr#; v# - sv#; count

sr# = r#

a$ = INKEY$

INPUT a$

IF a$ = "s" THEN STOP

test = v# >= (r# + 99)

IF test = 0 THEN z = 0

IF test = -1 THEN z = 1

IF test = -1 THEN CALL catch(v#, r#)

delta = (count > 25) AND (test = -1)

IF delta = -1 THEN z = 2

IF delta = -1 THEN count = 0

LOOP
```

```
SUB catch (v#, r#)

DO

r# = r# + 3

LOOP UNTIL v# - r# < 10
```

```
END SUB
```

```
"C:\Users\Reactor1967\vcns\work\012404F.BAS"

REM This did not work as I entended at this time in this configuration

REM but being without anyway to analize(misspelled) anything who knows.

REM but it did not seem to work for the fact that I ment to jump v# around

REM around from one coding range to the other instead v# ended up incrementing

REM by one everytime. Its been my experiance that when you do that and

REM increment r# that some overlapping occurs. Also, I,ve been looking at

REM a method of correcting the speed of r# for the data been incoding

REM as I code. In math what ever you do on one side of an equation you

REM do on the oppitsit(Misspelled) side. Maybe a balancing act can be done.

REM on r# or v# or both as I code. Just throwing that in there.

REM It is possible to have only one r# for each v# but you will have to

REM increment v# before it gets out of a coding range into the next

REM coding range not afterwards. I believe that is what is wrong with

REM this program.

DECLARE SUB findr (y#)
```

```
v# = 1

r# = 1

RANDOMIZE TIMER

CLS

DO

y# = v#

CALL findr(y#)

r# = y#

z = INT(RND * 3) + 0

IF z = 0 THEN v# = v# + ((v# - r#) * 2) + 1

IF z = 1 THEN v# = v# + ((v# - r#) * 2) + 2

IF z = 2 THEN v# = v# + ((v# - r#) * 2) + 3

PRINT , , z; v#; r#; v# - r#

a$ = INKEY$

IF a$ = "s" THEN STOP

LOOP



SUB findr (y#)

REM y# = y# + 300:rem ------------------------

REM v# = 1

REM r# = 1

redo:

a1# = v#
```

```
c1# = v# + ((v# - r#) * 2) + 3
FOR count = 1 TO 300
x# = v# + ((v# - r#) * 2) + 3
v# = v# + 1
NEXT count
b1# = (v# - 1)
d1# = x#
e1# = r#
REM -------------------------------
v# = b1# + 1: REM ----------
REM v# = d1# + 1: REM ----------
REM v# = d1#: REM ----------
REM -------------------------------
r# = v#
REM PRINT a1#; b1#; c1#; d1#; e1#:rem ------------------
delta = (a1# < y#) AND (b1# > y#)
IF delta = -1 THEN y# = e1#
IF delta = -1 THEN EXIT SUB
IF delta = -1 THEN GOTO hey:
tango = (a1# = y#) AND (b1# > y#)
IF tango = -1 THEN y# = e1#
IF tango = -1 THEN EXIT SUB
IF tango = -1 THEN GOTO hey:
bravo = (a1# < y#) AND (b1# = y#)
```

```
IF bravo = -1 THEN y# = e1#

IF bravo = -1 THEN EXIT SUB

IF bravo = -1 THEN GOTO hey:

z$ = INKEY$

REM INPUT z$

IF z$ = "s" THEN STOP

REM IF b1# > y# THEN STOP

GOTO redo:

hey:

END SUB
```

www.ingramcontent.com/pod-product-compliance
Lightning Source LLC
Chambersburg PA
CBHW081426170526
45166CB00008B/2110
* 9 7 8 1 4 9 2 1 7 0 8 8 4 *